The Pilot's
Radio
Communications
Handbook

4th Edition

TAB
PRACTICAL
FLYING SERIES

The Pilot's
Radio
Communications
Handbook

4th Edition

Paul E. Illman

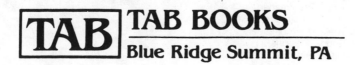

TAB BOOKS

Blue Ridge Summit, PA

FOURTH EDITION
FIRST PRINTING

Library of Congress Cataloging-in-Publication Data

Illman, Paul E.
   The pilot's radio communications handbook / by Pete Illman.—4th
 ed.
      p.   cm. — (Tab practical flying series)
   Includes index.
   ISBN 0-8306-4140-8 (pbk.)   ISBN 0-8306-4139-4
   1. Radio in aeronautics.   I. Title   II. Series.
 TL693.I4   1992
 629.132′51—dc20   92-2385
                                                      CIP

Acquisitions Editor: Jeff Worsinger
Editorial team: Christopher Cortright, Editor
               Susan D. Wahlman, Supervising Editor
               Joanne Slike, Executive Editor
               Jodi L. Tyler, Indexer
Production team: Katherine G. Brown, Director
               Lisa Mellott, Typesetting
               Donna Harlacher, Typesetting
               Olive Harmon, Typesetting
               Susan Hansford, Typesetting
               Tina Sourbeir, Typesetting
               Toya Warner, Layout
               Lorie White, Proofreader
               Linda King, Proofreader
               Joan Wieland, Proofreader
Design team: Jaclyn J. Boone, Designer
               Brian K. Allison, Associate Designer   PFS

Contents

About one
of the original authors

The initial concept for this radio communications handbook was the brainchild of Jay Pouzar back in the early 1980s. Relying on his extensive flying experience in the Coast Guard, in commercial operations, and as the owner of his own flight training school, he and I worked together to produce the first edition of the book that appeared in 1984.

While our paths had largely separated the past five years or so, it was with shock that I learned of his passing on December 1, 1991. Consequently, and as work on it had begun subsequent to his death, I am the sole author of this revised, updated, and partially rewritten fourth edition.

It is to Jay, however, that I owe my gratitude for conceiving the idea for such a book and for asking me to collaborate with him on the project. It was he who gave the project impetus; it was he who contributed so much of his own knowledge and expertise that enabled us to produce a well-accepted addition to the pilot's library.

Jay Pouzar will be missed by all who knew him and by those who have benefitted from his teachings—whether in person or via the written word.

Paul E. (Pete) Illman
March 1992

Acknowledgments

In the three previous editions of this book, we listed, with our sincere thanks, the many people who contributed so willingly of their time to assist us in the preparation of each edition. Those involved were almost exclusively associated with the Federal Aviation Administration, but many have since transferred to other posts, retired, or have perhaps severed their relationships with the FAA. Consequently, and with the hope that they will not be offended, we have chosen not to re-name those that were so helpful in the past. It should be noted, however, that they all held responsible positions in one of these facilities:

Olathe, Kansas, Air Route Traffic Control Center
Kansas City Downtown Airport Control Tower
Kansas City International Airport Tower and Approach/Departure Control
Kansas City Flight Standards Office
Wichita, Kansas, Approach Control/Departure
McAlester, Oklahoma, Automated Flight Service Station
Columbia, Missouri, Automated Flight Service Station
National Ocean Service
Washington, DC, Airport Control Tower
Kansas City FAA Regional Office.

In addition to those who previously gave so much of their time and expertise, it is important now to acknowledge with equal gratitude four individuals who contributed to this revised and updated fourth edition.

One is Dale Carnine, airspace and procedures specialist in the FAA's Kansas City Regional Office. Dale spent much time with me in person and over the phone briefing me on the airspace reclassification, the regulations pertaining thereto, and providing me with non-confidential written materials that became so valuable in the writing of certain chapters.

The second individual is Walt Roberts, Area Manager in the Columbia, Missouri, Automated Flight Service Station. With the changes brought about by the FSS consolidation program, Walt's willingness to critique and review with me the Flight Service chapter for its accuracy was essential. Additionally, I'm grateful to him for his always-patient responses to my calls when questions of doubt or uncertainty arose.

Finally, I am indebted to J. D. Green and Dennis Whittaker, Plans and Procedures specialists in the Kansas City International Tower and Approach/Departure Control facility. They gave freely of their time and carefully answered my many questions, especially those related to current Approach Control procedures. They both also reviewed the draft of the Approach/Departure Control chapter to ensure its procedural accuracy, making several critical corrections in the process, and were always available for subsequent queries or requested assistance.

Without the help of these gentlemen and, indeed, the help of everyone I ever contacted in the FAA, putting together a book of this nature would have been an infinitely more challenging task. To them and to all, past and present, I acknowledge my debt with thanks.

Introduction

The air is filled today with pilots of all levels of experience, knowledge, and training. There are those who learned to fly at some uncontrolled single-strip airport and are still reluctant to venture too far from that uncomplicated harbor. There are those who mastered the private test at a busy controlled airport. There are the old-timers who never lost the flying bug but couldn't afford the luxury until their later years. And there are the pros: the airline captains, the executive pilots, the high-time instructors, the commercial pilots who work charters or are involved in some other money-making enterprises.

The list is hardly complete. Suffice it to say that even with exorbitant fuel and maintenance costs, there are a lot of us in the air—some good, some bad, some in between.

What's the common denominator we all share? Probably the critical one is safety—and all that safety implies. Just below safety may be the freedom that piloting your own (or rented) aircraft brings: the freedom to go places with reasonable economy and speed, and the freedom to get from here to there unencumbered by traffic lights, speed traps, and highway nuts who pass you at 85 on a 65-mph interstate. Only the pilot enjoys the freedoms of space, distance, and speed.

A question, though, is the extent to which we take advantage of the benefits that the Cessna, the Cherokee, the Bonanza offer us. Said another way, how many of us find ourselves limited to the local traffic pattern or a few short hops to an uncontrolled airport because the big ones scare us? Or how many of us use—or know how to use— Approach Control, Center, or the Flight Service Stations? Or how many understand radio procedures and have mastered the techniques of pilot radio communications that make use of these various facilities possible?

Considering the emphasis placed on pilot training, medical qualifications, aircraft maintenance, operating rules and regulations, and the like, it would seem that at least a somewhat similar emphasis would be directed to pilot radio communication skills. For whatever reasons, such is not the case. The literature on the subject is too sparse, the examples of radio dialogue too few, and explanations of what to say and how to say it too incomplete.

Theoretically, a book on radio communications shouldn't be necessary; the subject should be part and parcel of every pilot's training. Innumerable discussions with controllers, airline pilots, instructors, and ordinary weekend excursionists, however, indicate just the opposite. According to the pros and amateurs alike, the airwaves suffer from communications misuse, non-use, or overuse. If such is the case, there can be little doubt that a void exists—a void either in the literature available, the pilot training process, or both.

In the effort to fill the void, this book has two primary purposes: to contribute to increased safety in flight through timely and correctly worded communications, and to equip the student and licensed pilot with the knowledge of radio communications and the various ground facilities so that his or her flight horizons are expanded beyond the local controlled or uncontrolled airport.

Designed primarily, but not exclusively, for the VFR pilot, the book discusses the whole spectrum of radio facilities and communication responsibilities. First, there is multicom, where only aircraft-to-aircraft self-announce messages are exchanged. Then comes a similar treatment of unicom, followed by Flight Service Stations, Ground Control, Tower, Approach/Departure Control, and the Air Route Traffic Control Centers.

Accompanying the discussion of the various facilities are explanations of what each does, how to determine the proper frequencies, and most important, examples of what the pilot should say to contact each facility, what he should expect to hear, and how he should respond.

This fourth edition, then, is an update of those that have preceded it, particularly in the areas of airspace reclassification, transponder and Mode C requirements, and the consolidation of Flight Service Stations into regional Automated Flight Service Stations (AFSSs). Also receiving more attention than previously are such matters as the proper radio communications procedures when an inflight emergency occurs, the sequence in which flight plan information is given and weather data reported when obtaining an AFSS weather briefing, the FAA Search and Rescue procedures, and pilot responsibilities in and around controlled airports.

In essence, the edition reflects revisions and updates that will hopefully make it a constructive addition to any pilot's library. More important, though, is the hope that it will contribute to the VFR pilot's storehouse of knowledge so that he or she can venture forth with greater confidence and greater safety.

Two Notes of Explanation

First, one would have to ignore reality if he were not aware of the number of women who are student pilots, licensed pilots, instructors, airline crew members, air traffic controllers, and so on. Their places in the world of aviation are steadily increasing, as are their contributions and their influence.

These being obvious facts, it's important to explain that, for ease of writing and the simplicity of sentence structure, I have generally used the male pronoun—he, him, himself—when referring to a pilot or a controller. The only reason for this choice is that the repetitive use of he and she, him and her, himself and herself, or any other dual gender combination is not only awkward but also can be disruptive to the reader. This explanation is offered here to preclude accusations of chauvinism or the failure to recognize the roles of women in the many fields of aviation.

Second, with a few exceptions, the aircraft N-numbers in the radio communication examples are those of planes that either Jay Pouzar or I once owned. Since the initial publication of this book, those aircraft have been sold to other parties—and perhaps resold. Consequently, should a subsequent owner find his N-number appearing in a communication example, let it not be a reflection of his radio skills or lack thereof. We chose our own N-numbers to avoid any such implication then, now, and in the future.

1
A Case for Communications Skills

The weather was clear as Jay Pouzar was returning to the Kansas City Downtown Airport from an IFR training flight. After Center had handed him off to Approach and he had established contact, another aircraft made its initial call, also to Approach:

Pilot: Clay County TCA Approach Control, this is Cherokee November Four One Nine Six Six. Over.

Approach: *Cherokee Four One Niner Six Six, Kansas City Approach.*

Pilot: Clay County TCA Approach Control, November Four One Nine Six Six is over the interstate, and I want to land at the big airport. Over.

Approach: *Cherokee Niner Six Six, squawk zero two five two, ident, and stand by.*

Pilot: Clay County TCA Approach Control, November Four One Nine Six Six squawking zero two five two, identing, and standing by. Over.

At this point the controller directed several other aircraft and lined Jay up for the instrument approach to Downtown. The controller then returned to N41966.

Approach: *Cherokee Niner Six Six, I missed your ident. Please ident again.*

Pilot: Clay County TCA Approach Control, November Four One Nine Six Six squawking zero two five two and identing. Over.

Approach: [After a pause] *Cherokee Niner Six Six, I'm still not receiving your ident. Remain clear of the TCA, and say your present position and altitude.*

Pilot: Clay County TCA Approach Control, I'm still over the interstate at three thousand five hundred feet, and I want to land at the big Kansas City Airport. Over.

Approach: *Cherokee Niner Six Six, which interstate are you over? There are several in this area.*

Pilot: Clay County TCA Approach Control, November Four One Nine Six Six. I'm not sure which interstate, but it's near the city. I still want to land at the big airport. Over.

Approach: *Cherokee Niner Six Six, I have not received your ident. Remain clear of the TCA and stand by.*

Instead of doing what he was told, the pilot of 966 launched into an airwave-monopolizing discourse along these lines:

Pilot: Clay County TCA Approach Control, this is November Four One Nine Six Six. I don't know why you aren't receiving my ident. I just had it worked on, and the mechanic told me it was fine. I've got to land at the big airport because I told Agnes, my wife, I'd pick her and the kids up when they got in from Chicago. What will they think if I'm not there? Over.

Approach: *Cherokee Niner Six Six, Kansas City International is a TCA, and I can't clear you to land unless your transponder is working. I am not receiving your ident, so remain clear of the TCA, and please stand by.*

Continuing to ignore the explicit instructions, 966 rambled on:

Pilot: Clay County TCA Approach Control, November Four One Nine Six Six. I just had the transponder checked because the last time I was here the controller told me to stand by. I did, and the thing didn't work then. The guy at the radio shop said it worked fine, but I'm still having the same trouble. Can't you get me into the big airport? Over.

Throughout all of this, other aircraft were trying to get a word in to report positions, get clearances, and the like. But N41966 continued on and on as though he was the only one in the air.

After the last exchange, the controller saw the light. The pilot of 966, bathed in the glow of ignorance, did what he had been told. He entered "0252" in the transponder, pushed the IDENT button, and placed the switch in the STANDBY position. Of course he wasn't received!

Once the mystery was solved, the pilot, quite unabashed by his display of incompetence, was cleared into Kansas City International—the "big airport."

This is about as accurate an account of the dialogue as is possible to re-create because, of course, Pouzar didn't tape the real thing. It's only one incident, and, while unusual in some respects, it's not very different from what pilots and controllers hear every day. All a pilot has to do is listen with a critical ear. Some of the garbage that filters through speaker or headset from air to ground reflects an appalling lack of knowledge that is both unfunny and potentially hazardous to the ignorant pilot and those occupying the same general airspace.

WHY THE PROBLEM?

Who's to blame for the incompetence? Oh, we could probably point a finger at the instructor who eased over the whole subject of communications, but the main thrust of accusation must be directed at the pilot himself. The pilot of N41966 obviously had little interest in the subject. Otherwise, he would have been sure that he knew what he was doing before venturing into a controlled and congested traffic area. At the same time, we can blame him for a consummate egotism that allowed him to enter such an area with so little knowledge.

Pilots such as our friend in N41966 are dangerous because they don't know what they don't know. They are the airman's example of the Peter Principle. They've risen to their level of incompetence. If the flying ability of the gentleman in the left seat of 966 is comparable to his communicating skills, he'll soon be soaring heavenward on his own wings—with his reliable little Cherokee rusting in some cluttered junkyard.

Yes, we can blame the pilot for incompetence, but others also share in the blame. Let's include the CFIs and CFIIs. And let's include the literature—or lack of it—that discusses the subject of radio communications.

Without exaggeration, it's a subject that probably receives the least attention and explanation of all of a pilot's flight training. Even the material currently in print offers little guidance on what to say and how to say it. The *Airman's Information Manual* takes a stab at a few examples, but the examples are limited; some conclude with the almost extinct "Over."

If flight instructors (certainly not all, but entirely too many) fail to teach more than the absolute rudiments of radio procedures, there are probably three reasons:

- The instructor isn't too certain about them himself. This should be an unlikely reason, but one of the original authors of this book, when getting back into flying after nearly 35 years as a groundling, was told by a young CFI to say "This is November 1111 Uniform" and always conclude with "Over." (November should really only be used by controllers when they don't know the type of aircraft. You should use Cessna, Cherokee, Bonanza, and so on, in your call sign. "Over" is rarely used, but nevertheless can be useful to indicate the end of a long transmission. Otherwise, it's not necessary.)

  If the instructor has accumulated most of his hours flying out of Cowslip Municipal, he probably isn't very confident of radio techniques. His own insecurity results in a superficial coverage of the subject as he preps his eager students for the FAA check rides.

- The instructor is a pro as a radio communicator, but teaching the subject takes time—mostly ground time, which is neither profitable nor exciting. So the student learns barely enough to get by.

- The airport is uncontrolled, with perhaps only unicom, and no controlled airport is within reasonable flight range. This is a logical reason for not teaching communications in depth, especially if the student plans to fly only on weekends and demonstrate his skills over Aunt Martha's barnyard. A good instructor emphasizes what is necessary to know, not what's nice to know.

    That same instructor, however, must make it very clear that if the student (now private pilot) ever plans to fly to a controlled airport or through a TCA (Terminal Control Area) or go on a cross-country, he must return for a thorough schooling in radio procedures. A fully equipped aircraft and a private pilot license entitle a pilot to land at any TCA airport. However, the hardware and a piece of paper are hardly adequate to ensure the continued well-being of the pilot or the other airmen in his vicinity. It's dangerous to run out of knowledge—but it can happen easily and quickly to untrained pilots. The results can be devastating.

So okay—you've got a license, and you either own or rent a plane. You're a good pilot, confident of your ability. Now, like many of your counterparts, are you going to spend the rest of your flying days avoiding tower-controlled airports or being fearful of using Center (Air Route Traffic Control Center) on a VFR cross-country? If you've been well-trained in radio procedures, a busy airport or getting advisories from Center is neither a challenge nor a concern. Your radio skill makes flying just that much more fun. But if you're untrained or uncertain, you'll probably steer clear of the controlled areas and not bother Center because you think that's for the IFR pilots and the pros who wheel the wide-bodies.

This, of course, is nonsense. Admittedly, Center might not be able to help you on a busy day if you're VFR. A controller also has the right to refuse to give you routine enroute advisories or track you on radar if you come across as hesitant, uncertain, or lacking in knowledge. Center controllers might not do this very often, but requests in VFR conditions have been rejected when the pilot was obviously incompetent in the basics of radio communications. Otherwise, Center exists to serve all pilots—from the greenest student to the 30,000-hour airline captain. Besides, the FAA urges us to use this as well as all other facilities available to us.

Let's be careful not to oversimplify the matter of radio procedures. Mastering them takes time and practice. To underscore that point, the FAA's *Instrument Flying Handbook* makes these comments:

> . . . Many students have no serious difficulty in learning basic aircraft control and radio navigation, but stumble through even the simplest radio communications. During the initial phase of training in Air Traffic Control procedures and radiotelephone techniques, some students experience difficulty. . . .
>
> . . . Communication is a two-way effort, and the controller expects you to work toward the same level of competence that he strives to achieve. Tape recordings comparing transmissions by professional pilots and inexperienced or inadequately trained

general aviation pilots illustrate the need for effective radiotelephone technique. In a typical instance, an airline pilot reported his position in 5 seconds whereas a private pilot reporting over the same fix took 4 minutes to transmit essentially the same information. . . . The novice forgot to tune his radio properly before transmitting, interrupted other transmissions, repeated unnecessary data, forgot other essential information, requested instructions repeatedly, and created the general impression of cockpit disorganization. . . .

PRACTICING FOR COMPETENCE

Mastery of the technique starts with knowing what you want to say, what to listen for, how to respond, and when and how to use the mike that spreads your voice throughout the surrounding skies. As in any other field, the initial ingredient of proficiency is knowledge. The trick is to apply that knowledge in a logical sequence so that you can say what you want to say and get off the air. Once the knowledge is acquired, the next step is practice, followed by more practice, until what you know intellectually becomes an ingrained habit.

If you've ever been asked to make a speech, you know that you didn't just get up and talk. You either wrote the entire speech or outlined it, and then practiced it until you had the subject matter, sequence, body language, and voice inflections down pat. The first time around, you were probably a bit nervous. The second time was a little easier. Eventually, if you spoke or lectured enough, you became an old pro.

It's the same thing talking to the ground from an airplane. Knowledge coupled with practice will calm your nerves and conquer whatever mike fright you might have. No matter how green you are, you'll come across as a professional.

You can practice in a couple of ways. One is to buy an inexpensive aircraft-band radio that picks up the various aviation frequencies. Then monitor the transmissions from your home. This method is less effective if you live far from a tower, but you can at least listen to the pilot's side of the communications exchange.

Another practice method is to use a tape recorder and do a little role-playing with yourself. You're the pilot and controller all in one. Make the initial call to Ground Control, and answer yourself as the controller would. Or pretend that you're in flight and want to land at X airport. Go through the same process. Using your knowledge of radio procedures, act out a series of scenarios on a mythical flight from the first contact with Ground Control until you have "landed," are off the active, and have called Ground Control again for taxi clearance.

Then play the tape back. Be your own worst critic. Be objective about the "dialogue." Ask yourself: "If I were a controller or another pilot listening to me, what would be my impressions of me?" If you're not satisfied, pick up the mike and go at it again.

If you practice this way enough, it won't take long to learn how to get the message across in the fewest possible words and with maximum clarity. Yes, you might feel a little silly sitting there talking to yourself, but that, too, shall pass. Even if it doesn't, it's a small price to pay for greater confidence and increased expertise.

Now you have the words down, but will you remember them, and in the proper sequence, when it comes to the real thing? If in doubt, write out what you want to say when you contact the various services. Put the notes on your knee pad and, if necessary, read from them as you make your calls. After all, people use checklists so they don't have to rely on a memory that might fail them, so why not adopt the same technique for radio contacts? (However, unlike checklists, experience makes the need for written notes unnecessary.)

A pause to regroup: Am I exaggerating the case and the need for greater communication skills among the pilot population? Obviously, I don't think so. All you have to do is fly a few hours a week and keep an alert ear to what flows through the headset. On any given flight around a busy airport you'll hear everything from a terse "Okay," to a rambling recitation of superfluous trivia, to a series of mumbled incoherencies that no human or electronic decoder could decipher. If you question that statement, spend a few minutes with a controller and listen to what he has to say.

CONTROLLERS ARE HUMAN, TOO

Flight Service Stations, Towers, Approach and Departure Control, and the Air Route Traffic Control Centers are the pilot's valuable but unseen friends. They exist to serve the pilot and to make flying safer for all. Their services, however, aren't really free. You've paid for them through your taxes. The services are there to be used or not used, so why not take advantage of your annual donation to Uncle Sam and the airway tax you find tagged onto your fuel bill?

Yes, you've paid for the service, but so has every other pilot, so it's not yours and yours alone to use or misuse. Any given service, particularly that offered by Center, can be denied you if you give an impression of incompetence. On occasion, those on the ground just don't have time to try to make sense out of nonsense and clarity out of obscurity. To do so might put someone else's life in jeopardy.

At times it might seem unlikely, but controllers happen to be human beings too. They have good days; they have bad days. Even on their good days, though, they can quickly turn into vocal ogres when they encounter unmitigated stupidity over the air. On bad days, they can come across as halo-endowed saints when a knowledgeable pro solicits their assistance or advice.

While we're on the subject, the basic rules of courtesy over the air should always prevail. When either the pilot or controller resorts to sarcasm or needless abuse, he is merely reflecting his emotional immaturity—which is hardly a credit to the responsibility he bears. Controllers call it "chipping," a gentle term for "telling the other guy off."

Controllers recognize the humanity of man, and 99 percent never utter a word of recrimination when mistakes are made or ignorance shines brightly. A few don't have that level of patience, of course. They can chip with the best of them, as did one I overheard when he couldn't get a response from a pilot with whom he had just been in contact: "You gonna talk to me, boy? If you are, talk *now*."

To give them their due, controllers have to be models of tolerance and self-control to endure some of the things that go on in and over the air. Yes, some talk too rapidly, and some runtheirwordstogether so that comprehension is nigh impossible. But the performance of the vast majority, even under pressure, sets a standard of excellence in their profession that pilots should strive to match in theirs.

If you have a problem with a controller, don't let anger overrule good judgment. The radio isn't the place for chipping. Wait until you're on the ground. Then call the facility and talk to the supervisor. Explain calmly what happened. Let the supervisor take it from there. Childish spleen-venting is out of place in the adult world, whether airborne or ground-bound.

While the controller is indeed the "controller," that doesn't mean he has to be obeyed at all costs. You are still the pilot in command. If the controller tells you to do something that you believe might endanger you, tell him. Don't follow him blindly into the path of possible destruction, but don't keep him in the dark about your concern or your alternate action.

In a very literal sense, a team is at work: you and the person on the ground. He is there to ensure your safety and that of your fellow pilots. He can fulfill his responsibility, however, only if you keep him informed and conduct yourself with the skill expected of a licensed pilot—private or ATP.

By the same token, if you help the controller when he asks you to lengthen your downwind leg, make a tight pattern, land long, speed up, slow down, make a high-speed landing runout, or whatever, you'll be functioning as an effective team member. Remember: the controller can do without you, but you can't do without him. Whether you are new or experienced, it is entirely to your personal benefit to make it easy for the controller to do his job and thus help you do yours. Achieving that objective is a matter of communications—knowing *what* to say, *how* to say it, *when* to say it, and *why* it should be said. Knowledge plus skill—the two added together equal professionalism.

All evidence that I have found indicates that a strong case can be made for greater pilot communication skills. The reason behind poor communication, whether it's the absence of literature on the subject or instructor reluctance to emphasize it, is secondary. The result is often a pilot's unnecessary fear of the microphone, which in turn tends to restrict his flying activities and limits the airborne adventures to which his license entitles him. The alternate result is unjustified confidence, as embodied by our friend in Cherokee November 41966.

What follows, from multicom to Center, will hopefully reduce your fear and establish justified confidence in your ability to communicate as a professional. Whether flying is your vocation or avocation, that should be your objective.

A FEW WORDS ABOUT PHRASEOLOGY

Because I'm about to begin illustrating the various radio calls and contacts, I want to be sure that the accepted phraseology is understood. There's nothing difficult about it, but there is a certain standardization that is both accepted and expected. Reasonable variations are, of course, permissible. The examples that follow in this book, however, generally reflect the approved wording and structure as established by the *Airman's Information Manual* (AIM) and the FAA's *Air Traffic Control Manual*, 7110.65, for controllers.

As to wording, aircraft N-numbers are stated individually, preceded by the aircraft type. Cherokee 1461 Tango is announced as "Cherokee One Four Six One Tango," not "Cherokee Fourteen Sixty-one Tango." Land Runway 19 is "Land Runway One Niner," not "Nineteen." "Altimeter 29.65" is "Altimeter two niner six five," "not twenty-nine sixty-five." "Heading 270" is "Heading two seven zero," not ". . . two seventy."

In quoting altitudes, controllers state them in terms of thousands and hundreds: "maintain three thousand five hundred" or "expect seven thousand five hundred in ten minutes." Pilots can (and do) shorten altitude quotations by saying "level at three point five" or "leaving five point five for three point zero" or "over the field at two point three." This sort of verbal shorthand is acceptable from the pilot, but does not conform to FAA standards. Consequently, you won't hear controllers using that phraseology, and in the examples I cite from now on, I attempt to conform to FAA recommendations.

Accordingly, and to be sure that you understand and employ the correct phraseology, all numbers in the simulated dialogues that follow are spelled out. "Runway 19" will appear as "Runway One Niner" because that's the way it's pronounced. "Heading 240" is stated as "Heading two four zero" and so on.

Depending on the specific reference, decimal points might or might not be included in the quotation. For instance, when citing altimeter settings, the decimal is omitted. A setting of "30.08" is communicated as "altimeter three zero zero eight." On the other hand, the decimal is included in references to radio frequencies. "Contact Ground on 121.9" is stated as "Contact Ground on one two one point niner" or "Contact Ground, point niner."

One other explanation is apropos before we get into examples. You will note that at times I use the aircraft's type and full N-number, such as "Cherokee One Four Six One Tango." On other occasions, it's "Cherokee Six One Tango." Why the difference? When making the initial contact with each controller (Ground Control, Tower, each Center sector, etc.), the type of aircraft should be identified and its full N-number given (just in case the controller is handling another "Cherokee Six One Tango," a distinct possibility in congested areas). You can shorten the call sign to "Cherokee Six One Tango" after the controller does. Once the controller abbreviates your call sign, there's no point in giving the complete identification in subsequent calls to the same controller.

Even at uncontrolled airports, the identification process should be the same. Make the type of aircraft you're flying known to others—and hope that they extend the same courtesy to you. There's a big difference between landing behind "Zero Zero Zero Zero Alpha" and "Learjet Zero Zero Zero Zero Alpha." It would be nice to know that you're trailing a jet rather than a Piper Cub. The wake turbulence of the former can be a bit more challenging.

Yes, a fair amount of verbal shorthand is acceptable, but not necessarily correct. Despite this phraseology latitude exercised by pilots (and perhaps even tolerated by the FAA), I have chosen not to take such liberties in the communication examples. The idea is to present the correct wording and phrasing here. Accepted but unapproved abbreviations can come later, if you so choose.

2
Multicom

A few years ago, I was driving with a friend down a well-traveled street in Riyadh, Saudi Arabia. Many side roads intersected this main street, but walls or buildings blocked the driver's view, making it impossible to see any cross traffic that might be approaching. Driving in Saudi Arabia is a thrill in itself, but when there are few stop signs or traffic lights and you can't see cars that might jump out at you from the left or the right, extreme caution is the only alternative for your continued physical well-being.

In this case, our driver friend slowed down at every blind intersection and honked his horn. At night, he honked the horn while blinking his lights. These signals alerted others that he was there. They were his nonverbal communication signifying his presence as well as his intentions.

In a more sophisticated sense, multicom is akin to the horn and lights of our Saudi driver. At airports with no ground-based traffic control or advisory service, no "red or green lights," no "stop signs," multicom provides the aural communications that reveal your presence and your intentions. Just like the Saudi driver, you are transmitting in the blind to anyone who is tuned to your frequency. You don't know if anyone is really listening or is even in the immediate vicinity, but like the driver, you take that added step—just in case.

In its simplest terms, multicom is nothing more than communication between two aircraft, whether in the traffic pattern or flying at altitude along the same route. While it can be comforting and perhaps important to talk to another pilot during a cross-country to exchange weather information and informal pilot reports (PIREPs), the real value of multicom comes to the fore around uncontrolled, nonunicom airports. That's when you need to know who is there, where he is, and what he intends to do—just as he needs to know the same about you. Multicom provides the vehicle for that exchange of information over a common frequency.

The key here is a common frequency. The FAA has thus established what it calls Common Traffic Advisory Frequencies (CTAFs). The CTAFs may be for airports that have no ground radio facilities at all, and are thus referred to as "multicom" airports, those that have only field-advisory radio services provided by the local Fixed Base Operator (unicom airports), Flight Service Stations, or tower-controlled airports that operate only part-time.

Two sources tell you what the CTAF is for a given airport. One is the *Airport/ Facility Directory (A/FD)*, as illustrated in FIG. 2-1. The other is the sectional chart (FIG. 2-2). If you refer to the sectional chart, you'll note the small circle with the letter "C", which indicates the CTAF for that airport. Immediately to the left of the circle is the frequency itself, in this case, 122.9. CTAFs that are printed in slanted, or italic, numbers and are magenta in color identify nontower airports, while the blue block-printed CTAFs, followed by the encircled "C," designate airports with control towers that operate part-time (as from 0700 to 1900). A full-time tower is indicated by the block-printed frequency, again in blue, but without the "C" symbol.

WAMEGO MUNI (69K) 3 E UTC-6(-5DT) 39°11′50″N96° 15′31″W   KANSAS CITY
966 B   L-6H
   RWY 17 – 35: H3170X30 (ASPH)   LIRL
   RWY 17: Thld dsplcd 170′. P-line.   **RWY35**: Trees.
   AIRPORT REMARKS: Unattended. Rwy lgts opr dusk – 0800Z‡. Rwy 17 – 35 3000X40′ lime chips with oil seal superimposed on
   3870X140′ turf. Rwy 17 – 35 lgts 35′ from Rwy edge. ACTIVATE LIRL Rwy 17 – 35—122.9 Ultralight activity on and in vicinity of arpt.
   COMMUNICATIONS: CTAF   122.9 ◄—
     WICHITA FSS   (ICT)TF 1-800-WX-BRIEF. NOTAM FILE ICT.
   RADIO AIDS TO NAVIGATION: NOTAM FILE MHK.
     MANHATTAN (T) VORW/DME   110.2   MHK   Chan 39   39°08′44″N 96°40′06″W   072.19°.4 NM to fld.   1060/9E.

Fig. 2-1. *The* Airport Facility Directory (A/FD) *identifies the 122.9 CTAF frequency at Wamego.*

Keep in mind these points about CTAFs: 1) "Common" doesn't mean one universal frequency for all airports, but rather the common frequency that all pilots should use for a given airport; 2) Despite what I just said, multicom—and only multicom—airports do have one common frequency nationwide—and that frequency is 122.9. Whatever the case, however, reference to the sectional or the *A/FD* is essential to determine the type of airport you are planning to enter and its correct radio frequency.

But back to operating in a multicom airport environment. Despite the importance

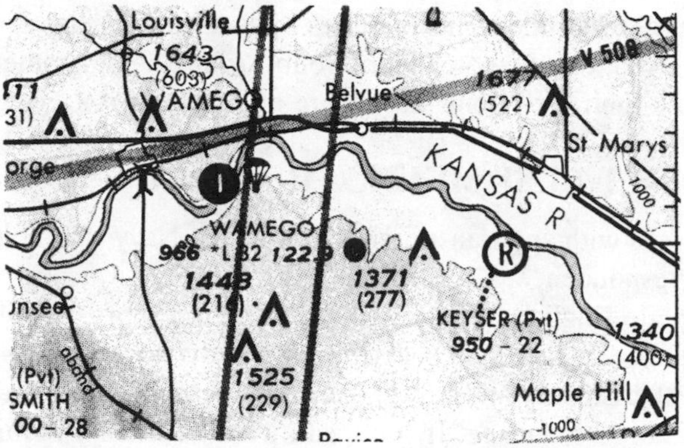

Fig. 2-2. *Wamego is also identified as a multicom airport on the sectional chart.*

of radio communications in such an environment, let's not be naive. No matter how clearly and explicitly you transmit your intentions, not all aircraft have radios. Even if they do, their pilots may not be tuned to the 122.9 frequency. Indeed, they might not have their radios on at all. ("Why bother about such things at Peapatch Municipal?")

The alternative is obvious. Flying around an uncontrolled airport demands a swivel neck and sharp eyes. To rely solely on blind transmissions is to invite a few thrills or unexpected encounters of the worst kind. Keep in mind, too, that your transmitter might suddenly go on vacation. You think you're broadcasting to all and sundry, but nothing is passing beyond the mouthpiece of your mike. Such failures have happened—and they could happen to you anytime, anywhere. Open eyes and constant head-turning are your best defenses against near hits or close misses.

WHY USE MULTICOM?

Why use multicom? The answer is already evident: safety. If you use it, other aircraft in the area might hear you and use it too. Then everyone will be informed about who is doing what, where, and when.

At the same time, don't be dumb. Be willing to back off when judgment so dictates. You've done a good job of advising others of your actions and intentions, but just as you're about to turn on final, you see some guy coming from nowhere on a long straight-in approach. Decision time—do you assume he knows you're there? Do you assume he'll give way because you're apparently number one to land? Go ahead and assume, but it's a dangerous practice.

Discretion says give way. In this case, it's better to be number two than arrive at the same point in a dead heat. Piggybacking might be the way to get a space shuttle to Cape Kennedy, but it's not a very comfortable way for a Cherokee and a Cessna to make a landing. That's a "dead heat" in a very literal sense.

A patently obvious observation? If it is so obvious, why do we still have midair collisions, landing accidents, and an excessive number of close calls? Part of the reason might be complacency ("It can't happen to me.") Part could be ignorance or stupidity—or a combination of the two. In some cases, it's nothing but a flagrant disregard for the rights of others. Whatever the reason, the diligent use of multicom at least reduces the potential of trouble while helping others who might be as concerned about their well-being as you are about yours.

HOW TO USE MULTICOM

As with any transmission, follow the Navy's admonition about report writing and correspondence: KISS—Keep It Simple, Stupid. Know what you're going to say, say it in plain English, say it clearly, and keep it short. If you've practiced your radio technique as suggested in chapter 1, you know how you sound. You should have learned to speak distinctly and slowly enough to be understood and to convey your message in an organized sequence. If you speak with the listener in mind, you'll communicate more effectively with fewer words and less monopolizing of the airwaves.

A SIMULATED LANDING AND DEPARTURE WITH MULTICOM

With the "theory" out of the way, let's go through the multicom procedures when operating into and out of an uncontrolled airport. Some of the transmissions might seem repetitious, but keep in mind that safety is the first concern. A single transmission might be the one that saves the ship.

Approach to the Field

Tune to 122.9 about 10 miles out. If there's any activity, you might get an idea of the traffic volume and pick up the favored runway and wind direction. Assuming this to be the case, start the self-announcing process by identifying your aircraft, position, altitude, and intentions. As there is no control agency of any sort on the field, your call is designed to advise other aircraft of your presence in the area. Consequently, open the transmission with the name of the airport, followed by "Traffic." Let's assume that you're going into Wamego (Kansas) Municipal (FIGS. 2-1 and 2-2), with its elevation of 966 feet and a 17−35 runway. Let's further assume that you've heard at least one other aircraft report its position in the traffic pattern and have learned that Runway 17 is the runway in use. The call, then, would go like this:

> Wamego Traffic, Cherokee One Four Six One Tango is ten north at five thousand five hundred. Will enter left downwind for full stop One Seven Wamego.

Note that the name of the airport is repeated at the end of the transmission. This is to make certain that there is no confusion on the part of other aircraft as to the airport to which you are going or at which you are operating. This positive identification is especially important when two or three other airports are in the general vicinity, each with several aircraft in the air, and all transmitting on 122.9. Identifying the intended airport twice will minimize the potential for confusion.

Going back to the arrival example, if you hear nothing after tuning to 122.9, don't be lulled into the belief that the skies are clear. Give yourself every safety edge you can. Announce your intentions to those who might be listening:

Wamego Traffic, Cherokee One Four Six One Tango is ten north at five thousand five hundred. Will cross midfield at two thousand five hundred for wind tee check and landing Wamego.

Over the Field

Assuming that you've heard no other traffic, you're over the field and see that the wind tee or sock favors Runway 17. Again, announce your intentions to the seen or unseen audience:

Wamego Traffic, Cherokee Six One Tango over the field at two thousand five hundred. Will enter left downwind for Runway One Seven, full stop, Wamego.

Entry to Downwind

You're entering the downwind leg at pattern altitude—approximately 800 feet agl. Get on the air again:

Wamego Traffic, Cherokee Six One Tango entering left downwind for Runway One Seven, full stop, Wamego.

Turning Base

Going into the turn from downwind, make your next call:

Wamego Traffic, Cherokee Six One Tango turning left base for Runway One Seven, full stop, Wamego.

At any uncontrolled airport, make this call when turning *onto* the base leg. It's a lot easier for other aircraft to see you when you're in a bank as opposed to straight-and-level flight. Also, a fairly wide pattern with a distinct base leg is preferable to a hotshot U-turn. This gives you time to scan the area for other aircraft on the same leg that haven't announced their presence. It also allows you to check the final approach course for someone who might be making a straight-in approach. This is the altar on which many inflight marriages have been consummated—unwanted but nevertheless eternal marriages.

Turning Final

Once again, announce what you're doing while in the turn to final:

Wamego Traffic, Cherokee Six One Tango turning final for Runway One Seven, full stop, Wamego.

Clear of the Active

On the ground, get off the runway as quickly but safely as possible. Somebody you never heard of might be on your tail. When clear, don't keep it a secret:

Wamego Traffic, Cherokee Six One Tango clear of One Seven, Wamego.

A lot of talk? Yes, but at least you've fulfilled your responsibilities. You've kept others informed, and you've made the air just that much safer. Now if only somebody as considerate is listening. . . .

You've gassed up, coffeed up, and are ready to go again. Once the engine has fired and the radio is on, listen for a moment or two while you're still on the ramp to see if any traffic has developed. As in the prelanding, your actions can be planned according to what you hear—or don't hear. Regardless, don't assume! Announce your intentions. At most one-strip fields a taxiway is an unknown luxury. Back-taxiing on the active is the only way to get into takeoff position. But let the other guy, if he's out there, know what you're going to do.

Taxi and Back-Taxi

Either while stationary on the ramp or moving slowly toward the runway, make your initial call:

> Wamego Traffic, Cherokee One Four Six One Tango on the ramp, will be back-taxi-
> ing on Runway One Seven, Wamego.

Before venturing onto 17, slow down and scan the approaches for both 17 and 35. It's possible that some unheard-from individual is landing downwind or is indeed on the final for 35 but hasn't had the courtesy to inform you.

If the air is clear, start the back-taxi, but don't dawdle. Apply some power and get to the end as rapidly and safely as you can. Otherwise, an approaching aircraft might have to go around. Or worse yet, he might not see you in time to abort his landing— and that could develop into a rather messy situation.

Preflight Runup

Let's assume that there is an area at the end of the runway for the preflight engine runup. (If there's not—and you should know before leaving the ramp or the taxi strip to the runway—complete the check before back-taxiing. Then, when you're at the end, all you have to do is turn around and go.) If an area is provided, do a 180 at the end, and park on the right side of the departure runway, if you're in the left seat. Then, after the pretakeoff check has been completed, you can turn toward the runway at a 90-degree angle and have a clear view of any activity on the final approach. Checking for traffic isn't easy if you're in the left seat and are taking the active from the left side of the runway.

This might be a small point, but I've had more than one pilot pull onto the runway when I was on final. They simply hadn't seen me, and a go-around was the only alternative. Was he using multicom? That's a silly question.

Taking the Active

You're ready to go, with the aircraft at an approximately 90-degree angle to the runway. Stop, look upwind and downwind (at this point some pilots swing a complete

360 to observe the entire traffic pattern), and then make your call:

> Wamego Traffic, Cherokee One Four Six One Tango departing Runway One Seven, Wamego.

On the matter of someone landing downwind, four conditions could bring about such a landing: 1) very light wind; 2) the pilot's complete disregard of the wind tee or sock; 3) the pilot's failure to monitor or use multicom; or 4) a genuine emergency. All but the fourth are caused by the third. With observant eyes and the radio tuned to 122.9, there is no excuse for aircraft landing in opposite directions at an uncontrolled airport. And yet, because of pilot ignorance, lack of radio equipment, or failure to use the equipment, this sort of thing happens too frequently.

To wit: I recently saw an individual barrel-in downwind, narrowly missing a landing plane that was doing everything correctly. He discharged his passenger and then roared off from the taxiway-runway intersection, which left him about 2,500 feet of a 4,000-foot strip. This time, however, he went into the wind. Did this hotshot in his sleek Bonanza use the sophisticated radio equipment that the plethora of antennas implied? Not once. He was apparently above such trivialities.

Departure

Although you've already stated your flight intentions, it doesn't hurt to repeat them once airborne as a courtesy to those still in the pattern:

> Wamego Traffic, Cherokee Six One Tango taking Runway One Seven, departing the pattern to the east, Wamego.

Until you're about 10 miles out, stay tuned to 122.9 to pick up any traffic that might be inbound, in your line of flight, or at your altitude. Also, you could help an arriving pilot who has just made his initial call by giving him the wind and runway information. For example:

> Aircraft calling Wamego Traffic, Cherokee Six One Tango just departed Wamego. Favored runway is One Seven, winds about two five zero at six.

There's no set pattern for such a call, so help the other pilot in your own words. Use the word "favored," however, when giving runway information. "Active" implies that a particular runway must be used—which, at an uncontrolled airport, is not the case.

Touch-and-Gos

Let's suppose that instead of departing the pattern, you want to make a few touch-and-gos. If you're parked at the ramp, the calls are the same up to the time you're ready to take off. Then get on the air:

> Wamego Traffic, Cherokee One Four Six One Tango taking Runway One Seven, closed pattern for touch-and-gos, Wamego.

On the downwind, turning base, and turning final, repeat your intentions, just as you did for the full-stop landing:

> Wamego Traffic, Cherokee Six One Tango turning downwind for Runway One Seven, touch-and-go, Wamego.

> Wamego Traffic, Cherokee Six One Tango turning base for Runway One Seven, touch-and-go, Wamego.

> Wamego Traffic, Cherokee Six One Tango turning final for Runway One Seven, touch-and-go, Wamego.

Yes, that's a total of four messages for one touch-and-go, but you never know who has just entered the pattern. Your last message could be the first he has received. You can't be sure—so be safe.

When you've had enough for the day, you're going to land or leave the pattern. In either case, keep the other traffic informed. If it's the final landing, make the downwind, base, and final calls similar to those cited above, substituting "full stop" for "touch-and-go":

> Wamego Traffic, Cherokee Six One Tango turning downwind for Runway One Seven, full stop, Wamego.

If you're leaving the pattern, make the call after the last takeoff and when you have the aircraft safely under control:

> Wamego Traffic, Cherokee Six One Tango departing the pattern to the east, Wamego.

CONCLUSION

Inexperienced pilots; pilots who don't know how to maneuver in even moderate traffic; failure to give way to others; lack of knowledge of multicom uses and techniques; failure to turn on the radio; no radio at all: all are reasons for accidents at uncontrolled airports.

In many respects, the controlled airport, even with its five o'clock congestion, creates a greater feeling of security than setting down at a small-town, uncontrolled field. The forced and enforced radio communications make the difference. At hundreds of fields like Wamego, vigilance coupled with skillful use of multicom greatly enhances the safety and mental tranquility all pilots seek in flight.

3
Unicom

The most modest air-to-ground communication (and also one that can be very helpful) is provided by unicom. In a sense, it's a step up from multicom and a step below the tower communications in a controlled airport traffic area.

WHAT IS UNICOM?

Simply stated, unicom permits radio contact with a ground facility on the airport. At many locations without a tower or Flight Service Station, the unicom operator fills the void by giving "field advisories" to pilots who call in and request them. The advisory consists of wind direction and velocity, possibly altimeter setting, the favored runway, and any reported traffic. Unlike a tower, however, unicom is not a controlling agency. The operator gives information, and that's all. The rest is up to the pilot.

Unicom also provides services of a nonflight nature. For example, if you want a fuel truck available for a quick turnaround, the unicom operator can make the arrangements. Maybe you need a taxi, or you'd like the operator to call your office or home to advise someone of your arrival time. Unicom is there to help you.

Of course, if an operating Tower or FSS is on the field, unicom can't and won't give you runway, winds, or traffic information. That's the responsibility of the official facility. Unicom will, however, provide the nonflight services mentioned.

WHO OPERATES UNICOM?

At uncontrolled airports (the primary concern at the moment), the fixed-base operator (FBO) usually mans the unicom. The radio facility itself can be located anywhere at the airport, but it's generally in the lounge area where a call-in can be handled by the FBO manager or a jack-of-all-trades employee who answers the phone, keeps the books, and sells candy and sectionals. Typically, a barometer and wind speed/direction indicator are near the transceiver.

Keep in mind that the unicom operator is not a controller. Indeed, he might have only the most meager knowledge of what goes on in the air. He probably will give you the best information he has, but it's unwise to count on 100-percent reliability.

For example, the wind direction, its velocity, and the favored runway might be completely accurate. The operator then concludes with "no reported traffic." In reality, half a dozen planes might be in the pattern, but none has been using the unicom frequency for that particular airport. Admittedly, if six airplanes are flying around him, it's a bit farfetched to believe that he doesn't know that traffic is in the area; the point, however, is that there has been no radio contact with him or on his frequency. In effect, there's "no reported traffic."

Don't disbelieve the unicom operator, but learn not to depend on his every word. He simply might not be in a position to know everything that's going on outside. Perhaps he's not even a pilot. Or, as in every walk of life, he could be one of the few who just doesn't care. The moral is to use the service, but be vigilant as you enter the airport area. Your own eyes are the best instruments you have to tell you what's happening in the real world.

HOW DO YOU KNOW IF AN AIRPORT HAS UNICOM?

The easiest way is to check the sectional. Just to be sure there is no confusion about what that chart tells you in that regard, let's take a couple of examples.

Figure 3-1 shows an uncontrolled airport (identified on the chart by the magenta coloring of the airport symbol and the related data). The italicized numbers indicate the field elevation—869 feet msl, the length of the runway—65, or 6,500 feet, and the unicom frequency—122.8, followed by the "C" CTAF symbol. The "L" by the runway footage does not mean length, but that the runway is lighted. The star or asterisk "*" just preceding the "L" indicates that the lighting is available on request, is part-time lighting, or is pilot-controlled. The larger star at the top of the airport symbol signifies the existence of a sunset-to-sunrise rotating light beacon.

Going a step further, but still on the subject of unicom, FIG. 3-2 illustrates how the sectional depicts the Columbia Regional Airport, which is tower-controlled. (All symbols and data relating to such an airport are always colored blue on the sectional.) With or without coloring, the fact that a tower is on the field is established by the "CT" (Control Tower) and the tower frequency, in this case, "119.3." Next comes the small star reflecting that the tower operation is part-time, followed by the "C", or CTAF symbol. That symbol means that transmissions should be made on the tower

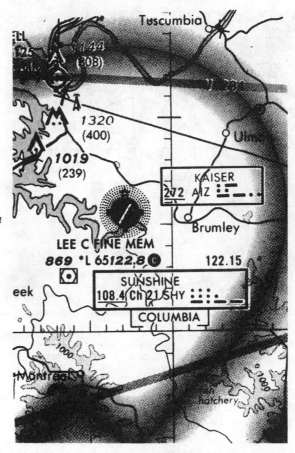

Fig. 3-1. *Lee C. Fine Memorial is a typical unicom airport, as depicted on the sectional.*

frequency of 119.3 even when the tower is closed. As to the other data, don't worry now about the "ATIS" reference or the "FSS" at the top of the data block (signifying the existence of a Flight Service Station on the field). Those will be discussed at the appropriate time.

Unicom service is also available at this airport, but the only indication of that is the italicized 122.95 frequency. In this instance, unicom could be contacted for whatever nonoperational services you might require, such as the need for a mechanic, a taxi, a call to the office, or the like. Otherwise, there's no reason to contact unicom because all flight-related information or instructions come from the tower controllers. When the tower is closed, the locally based FSS—not unicom—provides weather-related data, the favored runway, and reported traffic on the 119.3 CTAF frequency.

As with multicom, another source of airport data is the *Airport/Facility Directory*. Figure 3-3 illustrates what the *A/FD* says about Columbia Regional Airport.

But back to the uncontrolled airport with unicom. Unicom's major drawback is that it might not be in operation when you need it. Maybe it's closed; maybe the person in charge has gone to lunch, is gassing an airplane, is on the phone, or just doesn't

Fig. 3-2. *Columbia Regional is an example of a tower-controlled airport with a unicom facility for nonoperational services.*

COLUMBIA REGIONAL   (COU)   10 SE   UTC–6(–5DT)   38°49'05"N 92°13'10"W   **KANSAS CITY**
889   B   S4   FUEL 100LL, JET A   OX 2   ARFF Index A   **H-4G, L-21A**
RWY 02-20: H6500X150 (CONC-GRVD)   S-92, D-125, DT-215   HIRL   **IAP**
   RWY 02: MALSR.   RWY 20: ODALS. VASI(V4L)—GA 3.0° TCH 39'.
RWY 13-31: H4401X75 (ASPH)   S-12, D-16   MIRL
   RWY 13: REIL. VASI(V2L)—GA 3.0° TCH 44'. Road.   RWY 31: REIL. VASI(V2L)—GA 3.15° TCH 33'.
AIRPORT REMARKS: Attended Mon–Fri 1200–0600Z‡ and Sat–Sun 1300–0500Z‡. PPR for unscheduled air carrier
   operations with more than 30 passenger seats, call safety officer 314–443–2811. ARFF Index B level svc avbl on
   req. When twr closed ACTIVATE HIRL Rwy 02–20 MALSR Rwy 02 and ODALS Rwy 20—119.3. NOTE: See
   SPECIAL NOTICE–Simultaneous Operations on Intersecting Runways.
COMMUNICATIONS: CTAF 119.3   ATIS 128.45 (1300–0500Z‡)   **UNICOM 122.95** ⬅
   FSS (COU) on arpt 122.65 122.2 TF 1–800–WX–BRIEF. NOTAM FILE COU.
   APP/DEP CON 120.0 (1300–0500Z‡)   **KANSAS CITY CENTER APP/DEP CON** 118.4 (0500-1300Z‡)
   TOWER 119.3 (1300–0500Z‡)   GND CON 121.6
RADIO AIDS TO NAVIGATION: NOTAM FILE COU. VHF/DF ctc COLUMBIA FSS
   HALLSVILLE (L) VORTAC 114.2   HLV   Chan 89   39°06'49"N 92°07'41"W   188° 18.2 NM to fld. 920/6E
   (L) VOR/DME 110.2   COU   Chan 39   38°48'39"N 92°13'05"W   at fld. 883/3E. HIWAS.
   ZODIA NDB (LOM) 407   CO   38°43'00"N 92°16'06"W   018° 6.5 NM to fld. Unmonitored.
   ILS/DME 110.7 I-COU Chan 44 Rwy 02 LOM ZODIA NDB. ILS Unmonitored when twr clsd.

Fig. 3-3. *The* A/FD *confirms the unicom frequency of 122.95, but note that the CTAF for traffic is 119.3 when the tower is closed—not the unicom frequency.*

hear your call over the blare of rock music and the hangar talk of gathered pilots, or maybe he just doesn't want to be bothered. The vast majority of unicom operators are conscientious businesspeople who want to render a service; a few, however, merely consider unicom an interruption of more rewarding pursuits.

The service can thus be limited. As a vehicle for keeping others informed and aware of your presence in the vicinity, however, unicom provides a safety measure that every pilot can and should use.

CONTACTING UNICOM

At airports with unicom but no tower or FSS, the unicom frequency is almost always 122.7, 122.8. or 123.0. These are the frequencies for airport advisories at uncontrolled airports. Let's say, then, that you're going into the Lee C. Fine Memorial Airport illustrated in FIG. 3-1. The call sign, or correct radio address, for Fine is "Kaiser", so 10 or 15 miles out, you tune to 122.8 and listen. Just as you did with multicom, you monitor the frequency to see what you can learn from aircraft that might be in the pattern or what information Kaiser unicom might be relaying to others that have already called in. If you can pick up the winds, favored runway, and so on merely by eavesdropping, all the better. You can spare the airwaves that one transmission.

Let's assume, though, that all is silent as you approach Kaiser. At least 10 miles out, the initial call to obtain the field advisory should go like this:

You: Kaiser unicom, Cherokee One Four Six One Tango.

Unicom: *Cherokee One Four Six One Tango, Kaiser unicom.*

You: Kaiser, Cherokee Six One Tango is ten miles south at four thousand, landing Kaiser. Request field advisory.

Unicom: *Cherokee Six One Tango, wind two three zero at one zero, variable. Favored runway is Two One. Reported traffic two Cessnas in the pattern.*

You: Roger, Kaiser. Thank you. Six One Tango.

A word of caution here: If you're monitoring unicom or listening to other aircraft as they report their positions in the pattern, be sure you're getting the right information from the right field. To illustrate what we mean: In one area of Kansas, at least four uncontrolled airports are within 25 miles of each other, all with the same 122.8 CTAF. Obviously, you don't have to have much altitude to hear all four (plus a few others more distant), so you could be picking up the winds and traffic at Airports B, C, or D when you were intending to land at Airport A, or vice versa. That's why it's essential to begin and end your transmissions with the name of the airport, just as you did with multicom.

At this point, several options are open to you. Your only intention is to land and tie down. Then your response is simply "Roger, Kaiser. Thank you. Six One Tango."

But maybe you want a taxi. Then it's "Roger, Kaiser. Would you call us a taxi to take us to Zandu Products?"

Or you need a mechanic: "Roger, Kaiser. Would you have a mechanic available? Our oil temperature is running high."

Or you'd like unicom to make a phone call: "Roger, Kaiser. Would you call 555-5678 and advise Mr. Schwartz that his party will be landing in about 15 minutes?"

But what if unicom doesn't respond to your first call? Try to rouse the operator a couple of more times. If you still get no response, make a blind call on 122.8 to any aircraft that might be in the pattern:

"Any aircraft at Kaiser, this is Cherokee One Four Six One Tango. Can you give me a field advisory?"

If someone answers, fine. You can then proceed to enter the pattern. Otherwise, it would be smart to fly over the field for a wind tee or sock check. The call, then, is the same as with multicom:

Kaiser *Traffic*, Cherokee One Four Six One Tango is ten south at four thousand. Will cross the field at two thousand five hundred for wind check, landing Kaiser.

Note that Traffic in this example is italicized. That's for emphasis, because once you have a field advisory from unicom or by other means, the unicom operator is out of the picture. Consequently, all position reports and flight intentions are now directed to other aircraft in the pattern or in the vicinity. Thus the calls are addressed to "(Blank) Traffic," not "(Blank) unicom."

A SIMULATED LANDING AND DEPARTURE WITH UNICOM

With or without an initial unicom contact, all calls follow the models we illustrated in the multicom chapter. However, to be sure that the routine transmissions are clear, let's go through them again, but without repeating the various conditions and observations.

Before Entering the Pattern

Case 1:   You received no field advisory from any source, so you fly over the field to determine the wind direction:

Kaiser Traffic, Cherokee One Four Six One Tango over the field at two thousand five hundred. Will enter left downwind for Runway Two One, full stop, Kaiser.

Case 2:   You received the field advisory from unicom or another aircraft:

Kaiser Traffic, Cherokee One Four Six One Tango entering left downwind for Runway Two One, full stop, Kaiser.

An admonition I must repeat: Keep your eyes open. Yes, unicom might have said that two Cessnas were reported in the pattern, but do you know that they constitute the only traffic? Has someone else shown up with no radio, a radio that hasn't been turned on, or who hasn't bothered to report his presence? This is an uncontrolled airport, so your most effective life preserver might be healthy skepticism liberally sprinkled with vigilance.

Turning Base and Final

Kaiser Traffic, Cherokee Six One Tango turning left base for landing Two One, Kaiser.

Kaiser Traffic, Cherokee Six One Tango turning final, landing Two One, Kaiser.

Down and Clear of the Runway

Kaiser Traffic, Cherokee Six One Tango clear of Two One, Kaiser.

Predeparture

Leaving a unicom airport is essentially the same as with multicom, except that a call to unicom can give you the wind, favored runway, and reported traffic. This call assumes, of course, that you haven't obtained the information from a personal visit with the unicom operator—which might not always be practical because you could be at one location on the field while the FBO is somewhere else, perhaps hundreds of yards away. Assuming you haven't talked with the operator, the call is simply:

Kaiser unicom, Cherokee One Four Six One Tango at (state your location on the field). Request airport advisory.

Once you have the advisory, and while still stationary or taxiing slowly, get on the air again and tell the local traffic what you're doing—or going to do:

Kaiser Traffic, Cherokee One Four Six One Tango taxiing to (or back-taxiing on) Runway Two One, Kaiser.

Departure

Kaiser Traffic, Cherokee One Four Six One Tango taking Two One, departing to the east, Kaiser.

After Takeoff

Kaiser Traffic, Cherokee Six One Tango clear of the area to the east.

Touch-and-Gos

As with multicom, communicate your intentions before taking off:

Kaiser Traffic, Cherokee One Four Six One Tango taking Two One, closed pattern for touch-and-gos, Kaiser.

On downwind:

Kaiser Traffic, Cherokee Six One Tango turning downwind for Two One, touch-and-go, Kaiser.

Turning base:

Kaiser Traffic, Cherokee Six One Tango turning base for Two One, touch-and-go, Kaiser.

Turning final:

Kaiser Traffic, Cherokee Six One Tango turning final for Two One, touch-and-go, Kaiser.

Landing or Departing the Pattern After Touch-and-Gos

When you're finished practicing and want to land, advise the traffic accordingly on downwind, base, and final:

Kaiser Traffic, Cherokee Six One Tango downwind for Two One. Full stop, Kaiser.

And so on.

If departing the pattern, the following will keep other traffic informed:

Kaiser Traffic, Cherokee Six One Tango departing the area to the east, Kaiser.

And a few minutes later:

Kaiser Traffic, Cherokee Six One Tango clear of the area to the east, Kaiser.

Now stay tuned to the Kaiser CTAF until you're 10 miles or so from the airport. This will help alert you to possible inbound traffic that might be in your general line of flight. Also, if you happen to hear an inbound pilot calling Kaiser unicom for a field advisory, don't jump in and volunteer the information, as suggested in the multicom chapter. Let unicom provide the information. However, if unicom doesn't respond after a number of efforts, it's proper to help the other pilot, based on what you knew a few minutes ago:

Aircraft calling Kaiser unicom, Cherokee Six One Tango just departed Kaiser. Favored runway is Two One, winds about One Niner at two zero, Kaiser.

CONCLUSION

Flying around an uncontrolled airport presents many opportunities for unwanted confrontations. Whether multicom or unicom, the potential is the same. All I can do is urge caution and compliance with the standard radio procedures.

To quote the FAA:

. . . increased traffic at many uncontrolled airports require[s] the highest degree of vigilance on the part of pilots to see and avoid aircraft while operating to or from such airports. Pilots should stay alert at all times, anticipate the unexpected, use the published CTAF frequency, and follow recommended airport advisory practices.

Whether you are all by yourself in the pattern or one of several pilots, follow the radio procedures we've outlined. Coupled with a swivel neck and alert eyes, sharp radio technique makes the air a lot safer for everyone—especially you.

4
Flight Service Stations

As close as the phone—or, less likely these days, maybe a short walk—is the pilot's supermarket of information and assistance. No, the Flight Service Station can't do everything, but when it comes to flight planning, weather, airport advisories, or almost everything, the FSS is a storehouse of aids that pilots need, use, or should use. If you're merely going to shoot a few touch-and-gos at the local aerodrome or wander around within a 25-mile radius of home base, the need for FSS services might be limited. But a venture farther from home demands at least a brief phone call.

For example, you're going to a field 30 miles or so away. A gusty crosswind blows you off the runway on landing and you wipe out a gear or catch a wingtip. Did you check the weather before taking off, and was your N-number recorded? If not, you might have to foot the repair bill out of your own pocket. Some aircraft insurance policies are invalid unless you can substantiate that you had taken that simple preflight precaution. The FSS was there, but you didn't use it.

THE SERVICES OFFERED: In summary

An FSS can offer you:

- Weather information—local, enroute, and terminal, including sky conditions, winds, temperatures, dewpoints, icing, frontal activity, trends—you name it, the FSS has it.

- PIREPs (Pilot reports) of conditions not easily determined from available charts or data.

- Flight plans—filing, changing, extending, and closing.

- Airport advisories for the airport where the FSS is located (when there is no tower on the airport or the tower is closed).

- Airport information (excluding traffic information) for airports with no FSS, but having a Remote Communications Outlet (RCO) and a weather observer.

- Status of Restricted Areas and Military Operations Areas (MOAs).

- NOTAMs—Notices To Airmen regarding airport conditions, hazards, etc.

- Emergency assistance services—direction finding (DF) fixes, steers, and approaches; VOR and ADF orientations; and so on.

Add one more. A sense of security, if you want to call it that. Thirty minutes after your estimated time of arrival, if you haven't extended or closed your flight plan, the FSS is on the phone checking your whereabouts. It's comforting to know that somebody down there is watching you.

With the library of help and information the FSS can offer, let's start at the beginning and follow a rough sequence of how you might make use of the various services.

FSS CONSOLIDATION

With economy and improved service the intent, the FAA is in the process of closing many of the approximately 300 Flight Service Stations and consolidating them into 61 Automated Flight Service Stations (AFSSs). This move, expected to be completed by 1994, has met considerable opposition from general-aviation pilots. Alaskan pilots have been particularly vocal, along with those in the lower 48 where local weather characteristics are of such consequence that on-site—not remoted—FSSs are considered essential to aviation safety.

As the result of strong lobbying, spearheaded by the Aircraft Owners and Pilots Association (AOPA), the FAA has restudied many of the potential closings, concluding that 31 of the existing FSSs were located in "significant weather areas" and justified retention. These are currently classified as Auxiliary Flight Service Flight Stations (XFSSs). Additionally, 15 "supplemental weather service offices" were added to the list. These XFSSs have qualified weather observers on board and are thus able to summarize local weather conditions, give airport advisories, and open or close flight plans. For briefing purposes, however, since the XFSS is not automated, the specialist has to obtain enroute weather conditions from its area AFSS by telephone or "Direct User Access Terminal" (DUAT)—which is also available to qualified pilots.

So, at the time of this writing, approximately 46 of the original FSSs will remain open (almost half of which are in Alaska), along with the 61 consolidated facilities.

All told, that's about a third of the facilities that existed in the preconsolidation days. Keep tuned, though. The figures might change again.

What AFSSs Mean to the Pilot

The major advantage of an AFSS is its automation. Through it, the specialists have more data at their fingertips, the services have been expanded, and the pilot can obtain a wider range of information by accessing the automated menus available by telephone.

The disadvantages of consolidation? Well, a common one, especially in the early days of consolidation, is the difficulty in reaching a briefer. It was, and still is, a telephone overload, especially when IMC (Instrument Meteorological Conditions) exist. Also contributing to busy lines, according to AFSS specialists, is caller disorganization and unpreparedness—a subject I'll address in a moment.

A second disadvantage is the absence of computer terminals in the XFSSs. This means, other than for local weather, that in-person briefings are about what the pilot could obtain by calling the appropriate AFSS himself.

In sum, though, the pluses of consolidation, so say many pilots, outweigh the minuses. If there are no incoming phone delays, briefings are faster and more thorough, flight plan filing is almost instantaneous, and many more automated services are available through pilot menu accessing.

OBTAINING A PREFLIGHT BRIEFING

Inasmuch as the vast majority of FSS contacts are by phone, it might be helpful at this point to summarize the pilot's responsibilities when obtaining a weather briefing. The essence is, first, planning the flight, and then organizing the information the specialist will require to give you the briefing.

According to AFSS specialists, one of the primary reasons for telephone-answering delays is the pilot's unpreparedness, the disjointed sequence of information he gives, and his uncertainty of what he really needs to know. Combined, or even in isolation, these deficiencies only extend the telephone call and keep others in a "please hold" pattern. As in the air, don't ramble on and consume the briefer's valuable time, such as this character did:

FSS: *Good morning. Lake City Flight Service.*

Pilot: Hi. I'm trying to get to Mountain Town. How are things between here and there? [Let's say Mountain Town is 300 NM from Lake City.]

FSS: *When do you plan to leave, sir?*

Pilot: Oh, in about an hour—say ten o'clock.

FSS: *Will you be VFR?*

Pilot: That is Roger. [That shows he's got the lingo down.]

FSS: *And at what altitude do you intend to fly?*

Pilot: I guess that depends on what you tell me about the winds. How're they running?

FSS: *Well, let's see. At three thousand they're at two nine zero at two zero; six thousand, two four zero at three zero; nine thousand, two two zero at three five—generally southwest to west.*

Pilot: And what's the ceiling? CAVU all the way? [That's an I-know-what-I'm-talking-about acronym for "Ceiling and Visibility Unlimited."]

FSS: *No. Lake City is reporting three thousand scattered, eight thousand broken, Midpoint Junction is two thousand five hundred broken, and Mountain Town is four thousand overcast.*

Pilot: Kinda marginal VFR in spots, isn't it?

FSS: *Yes, sir.*

Pilot: What's the forecast for later today?

FSS: *About the same conditions until twenty-one hundred Zulu.*

Pilot: Twenty-one hundred Zulu—that's three o'clock tomorrow . . .

FSS: *No, sir, twenty-one hundred is fifteen hundred local; and local is three p.m. today.*

Pilot: Oh, yeah. I always get confused about whether to add or subtract. Well, what else can you tell me?

FSS: *. . . Excuse me, sir. I have another call. Stand by.*

If the specialist is lucky, a telephone failure will now ensue.

Types of Briefings

Three types of briefings are available: Outlook, which provides trends 6 hours or more in advance of the planned flight departure; Standard, the normal full-length briefing; and Abbreviated, usually a capsulized update of a previously obtained Standard briefing. Because it's the most common, what follows focuses on the Standard.

Preparing for the Briefing

First, go through the normal flight planning steps—route of flight, heading, probable ground speed, proposed flight altitude, and the rest.

Second, complete the entire FAA Flight Plan form.

Third, call the AFSS (or FSS). Once the briefer is on the line, the Flight Plan form now comes into play. Slowly and clearly enough so that the briefer can digest what you're saying, read off your entries in Blocks 1 through 10 on the form (FIG. 4-1). Then stop. While you're reading, the briefer is typing the information into his computer terminal and, in effect, recording your flight plan. If the weather then permits a "go" decision, the plan is already in the computer. Following the briefing,

Form Approved: OMB No. 2120-0026

U.S. DEPARTMENT OF TRANSPORTATION FEDERAL AVIATION ADMINISTRATION **FLIGHT PLAN**	(FAA USE ONLY) ☐ PILOT BRIEFING ☐ VNR ☐ STOPOVER		TIME STARTED	SPECIALIST INITIALS

1. TYPE ✓ VFR IFR DVFR	2. AIRCRAFT IDENTIFICATION N1461T	3. AIRCRAFT TYPE/ SPECIAL EQUIPMENT PA28/A	4. TRUE AIRSPEED 110 KTS	5. DEPARTURE POINT DALHART	6. DEPARTURE TIME PROPOSED (Z) 1550	ACTUAL (Z)	7. CRUISING ALTITUDE 7500

8. ROUTE OF FLIGHT

DALHART — V234 — WICHITA

9. DESTINATION (Name of airport and city) WICHITA MID-CONTINENT WICHITA, KS	10. EST. TIME ENROUTE HOURS 2 / MINUTES 45	11. REMARKS

12. FUEL ON BOARD HOURS 4 / MINUTES 30	13. ALTERNATE AIRPORT(S)	14. PILOT'S NAME, ADDRESS & TELEPHONE NUMBER & AIRCRAFT HOME BASE C.O. FARMER (806) 555-1234 100 PASTURE LANE DALHART, TX DALHART MUNICIPAL 17. DESTINATION CONTACT/TELEPHONE (OPTIONAL) JOHN ROGERS (316) 555-6789	15. NUMBER ABOARD 2

16. COLOR OF AIRCRAFT WHITE/RED STRIPES	CIVIL AIRCRAFT PILOTS. FAR Part 91 requires you file an IFR flight plan to operate under instrument flight rules in controlled airspace. Failure to file could result in a civil penalty not to exceed $1,000 for each violation (Section 901 of the Federal Aviation Act of 1958, as amended). Filing of a VFR flight plan is recommended as a good operating practice. See also Part 99 for requirements concerning DVFR flight plans.

FAA Form 7233-1 (8-82)   CLOSE VFR FLIGHT PLAN WITH ___WICHITA___ FSS ON ARRIVAL

Fig. 4-1. *An example of a properly completed VFR Flight Plan form. Blocks 1 through 10 should be read to the briefer at the start of the briefing.*

assuming the flight is still "go," the briefer will ask for the rest of the information on the form—Blocks 11 through 18.

Considering what the briefer is doing, be sure to communicate the Block 1 through 10 elements in the exact sequence you have recorded them on the Flight Plan form. In other words, in this order:

1. VFR or IFR
2. Aircraft N-Number
3. Aircraft Type/Special Equipment letter code (See AIM, Section 5)
4. True Airspeed
5. Departure Point
6. Proposed Departure Time
7. Cruising Altitude
8. Route of Flight
9. Destination
10. Estimated Time Enroute

Plus: Type of briefing desired: Standard, Abbreviated, or Outlook.

The Briefing Sequence

Once the briefer has the data in Blocks 1 through 10, he'll provide the following information in this sequence:

1. *Adverse Conditions*, if any;

2. *VFR Flight Not Recommended*: If conditions appear unfavorable for VFR flight, it's at this point that the briefer will suggest "VFR is not recommended." Remember, though, the briefer is not a controller and has no authority to approve or disapprove a flight. So the go/no-go decision is yours. The fact that VFR was not recommended, however, is noted on the Flight Plan form at the very top in the small box labeled VNR (VFR Not Recommended) and it's wise to heed the briefer's suggestion. He's the professional. Assuming, though, that weather is no problem, the briefing continues;

3. *Synopsis*: A brief summary of weather systems that might affect the proposed flight;

4. *Current Conditions*: A summary from all sources of weather conditions applicable to the flight;

5. *Enroute Forecast*: A summary of forecast conditions in the sequence of climb out, enroute, and descent;

6. *Destination Forecast*: The forecast of significant changes expected within one hour before and after the ETA;

7. *Winds Aloft*: The forecast winds at the proposed flight altitude and, if requested, temperatures at that altitude;

8. *NOTAMs (Notices To Airmen)*: Unpublished NOTAMs that could affect the flight;

9. *ATC Delays*: Any that might affect the proposed flight;

10. *Upon Request*: Additional information, such as active MOAs (Military Operations Areas), active MTRs (Military Training Routes), and the like, within 100 miles of the AFSS facility.

Suggestion: Jot down those italicized headings on a piece of paper as part of your briefing preparation. You'll know what's coming and in what order. It's then just a matter of filling in the spaces.

Now, if necessary, go back and replot your flight, based on winds or other conditions the briefer may have given you that would affect your initial plan. If the changes are minor, such as a little longer or shorter enroute time, just communicate the new estimated arrival time to the FSS when you open the flight plan. If major alterations are necessary (route deviations, unanticipated intermediate stops, or the like), a call back to the FSS would probably be in order. It's easier to make significant revisions by phone than over the radio.

Just one more thing: Unless you already know it, ask the briefer to tell you what FSS frequency to use when opening your flight plan. The reason for that will be apparent in the next section.

There's a lot more to the FSS or AFSS than I've indicated here—such as "Fast File," Flight Watch, Pilot Reports, enroute position reports, or FSS help when lost or disoriented. Some of these will come into the picture later. For now, though, the concentration has been on the briefing service and what you, the pilot, can do to take full advantage of what the FSS has to offer. Believe me, the facility is there to help, and help it will when the pilot on the other end of the phone organizes his material, knows what he wants, asks pertinent questions, and hangs up. It's that simple.

A FEW WORDS ABOUT FSS FREQUENCIES

Note: Just so there is no confusion, and unless there is a specific reason to do otherwise, I'll henceforth refer to Flight Service Stations simply as "FSSs"—not "AFSSs" or "XFSSs." While there are indeed differences between the two, many of the same radio principles apply to both. Also, "Flight Service Station" is the generic and accepted term to identify the facility.

But first, four points about radio calls to an FSS are necessary.

- The FSS Inflight Specialist (not the Briefer) with whom you would normally have radio communications can have up to 48 individual or duplicated frequencies from many remoted locations to monitor. These are all displayed on a lighted panel at his position, and next to each light is a switch by which the specialist accesses a particular frequency. Then, when a pilot calls the FSS, the panel light identifying the frequency he is using starts to blink. The only problem, from the specialist's point of view, is that the light continues to blink only while the pilot's mike button is depressed. Once the button is released, the blinking stops and the light goes out.

   The point is this: The specialist might have heard the radio call, but if his attention was directed elsewhere and he did not see the light while it was blinking, he wouldn't know which frequency to activate. Consequently, it's important that you include in the initial call both the frequency you're calling on and your location. The location or airport name is important because two or more geographically separated airports can be assigned the same frequency.

- The following frequencies never appear on the sectional chart because they are universal and permit access to any FSS:

   121.5—for emergency use only;
   122.0—"Enroute Flight Advisory Service" (EFAS): The discrete frequency for enroute weather updates and flight guidance;
   122.2—for routine communications.

- Beyond the unprinted "universal" frequencies, many others appear on the sectional, particularly those associated with Remote Communications Outlets and VORs (Very High Frequency Omnidirectional Ranges). Just one example is the VOR illustrated in FIG. 4-2.

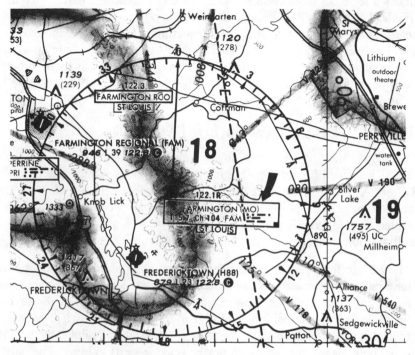

Fig. 4-2. *The Farmington, Missouri, VOR illustrates other frequencies that can be used to contact the FSS in St. Louis.*

The thin rectangular box identifies the VOR as "Farmington," which can be received in flight on 115.7 or Channel 104. Next is the continually transmitted identifying Morse code signal for "FAM." The small box in the lower right corner formerly indicated that the station provided continuous recorded Transcribed Weather Broadcasts (TWEBs) over a 400-mile radius of the VOR. TWEBs have been discontinued almost everywhere, however, and the square now means that HIWAS—Hazard Inflight Weather Advisory Service—is transmitted on a 24-hour basis.

Finally, the 122.1R above the rectangle is the frequency on which the AFSS receives transmissions. Thus, to contact the FSS over the VOR, the pilot would call on 122.1 and the FSS would respond on the VOR's 115.7 frequency.

- When contacting an FSS, use the word "Radio," not "Flight Service." Example: "Columbia Radio, Cherokee One Four Six One Tango . . . ." The only time "Radio" is not used is when you're calling Flight Watch. But here's another recent change: The call should be addressed to the Air Route Traffic Control Center AFSS in whose area the AFSS is located, as "Kansas City Flight Watch"—not "Columbia Flight Watch." Columbia still responds, so it's only the address that is different. This change is currently being made and will soon become systemwide.

OPENING THE FLIGHT PLAN BY RADIO

To open a flight plan with an FSS by radio, four situations can exist. The radio phraseology in each, however, is basically the same, with only minor adjustments, depending on the airport and the location of the FSS.

Situation 1: FSS is on the Departure Airport

You're at Hulman Regional in Terre Haute, Indiana (FIG. 4-3). The "FSS" just above the airport name and the heavy lined box to the left of the airport data indicate that the Terre Haute FSS (actually an AFSS) is located on the field and can be reached on 122.65. So, unless you were given a different frequency by the briefer, make the call to open the flight plan on 122.65.

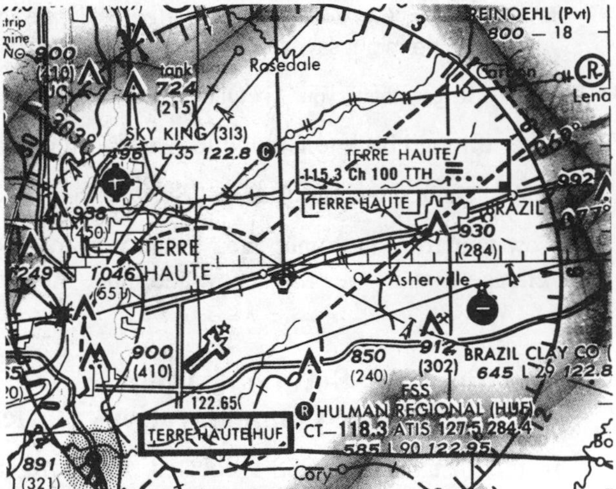

Fig. 4-3. *At Hulman, the Terre Haute AFSS is reached on 122.65.*

Next, when and how should the call be placed? Simple.

You've been cleared by Ground Control to taxi, let's say to Runway 3. You taxi out, come to a full stop, complete the preflight runup check, and are ready to open the flight plan. You're still under the jurisdiction of Ground Control (GC), however, so a call to Ground comes first to advise that you're temporarily leaving its frequency:

You: Hulman Ground, Cherokee One Four Six One Tango leaving you to go to Flight Service.

Be sure to wait for Ground's approval to change frequencies, and then switch to 122.65, or whatever frequency the briefer gave you.

You: Terre Haute Radio, Cherokee One Four Six One Tango on 122.65, Hulman.

Say nothing more now until the FSS has acknowledged your call:

FSS: *Cherokee One Four Six One Tango, Terre Haute Radio. Go ahead.*

You: Radio, Cherokee Six One Tango. Would you open my flight plan to Nashville at fifteen ten Zulu [or *". . . at one five one zero Zulu"*]?

Two points here: In filing the flight plan, you estimated a 10:00 a.m. EST departure, or 1500 when converted to the Greenwich, England, Coordinated Universal Time (abbreviated to "UTC" or "Z"—Zulu). UTC is thus the time to be used in the FSS contact.

Secondly, add about 10 minutes to the flight plan opening in case other aircraft are in line for takeoff clearance or there is a possible delay caused by incoming traffic.

Following the last call, Flight Service will come back with a response similar to this:

FSS: *Roger, Cherokee Six One Tango. Will open your flight plan at fifteen ten* [or he might say "ten past the hour"].

You: Roger, thank you. Six One Tango.

After this, go back to Ground to reestablish communications:

You: Hulman Ground, Cherokee Six One Tango back with you.

Now taxi from the runup area to the runway "hold" line. Come to a complete stop, change to the tower frequency, and make the call to the tower for takeoff clearance.

Situation 2: FSS Not on Airport but Has Remote Communications Outlet

In this instance, the FSS is located elsewhere, but a Remote Communications Outlet, or RCO, is on or close to your departure airport. As an example, you're at Rolla (Missouri) National (FIG. 4-4). The box to the lower left of the airport symbol indicates that the Columbia FSS can be reached on the ground (or in the air, for that matter) on the RCO 122.35 frequency.

Rolla being a unicom airport, you go through the usual ground traffic advisories on 123.0, as discussed in the previous chapter, and after engine runup, you switch to 122.35 for the FSS contact. The only difference of consequence here, versus the Terre

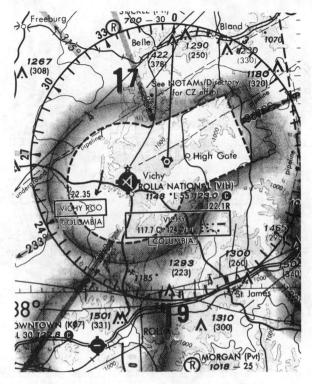

Fig. 4-4. *The small rectangle indicates the existence of an FSS RCO on or near the Rolla airport.*

Haute example, is in the initial call and the location identification:

You: Columbia Radio, Cherokee One Four Six One Tango on 122.35 at Vichy.
(Use "Vichy" rather than "Rolla" in this instance.)

The structure of the rest of the call is the same as in the Hulman example. Assuming, however, that local traffic permits an immediate takeoff, there's no need to add ten minutes to your departure time. If it's now 1500Z, just tell Flight Service to open your flight plan "at fifteen hundred Zulu."

Once the FSS contact is completed, go back to the 123.0 unicom frequency and continue with the standard reporting messages to Rolla traffic.

Situation 3: No FSS on Airport; No RCO; Airport Has an Adjacent VOR

You're on the ground at Vandalia, Illinois (FIG. 4-5). The FSS is in St. Louis, and there is no RCO on the field. The Vandalia VOR, however, is only 7 miles to the north and transmits on 114.3. At the same time, the "122.1R" above the VOR identifying box means that the FSS can receive on that frequency. What you do, then, is tune to 114.3 for VOR voice reception and to 122.1 on your radio. With this combination, you'll be transmitting on 122.1 and listening on 114.3.

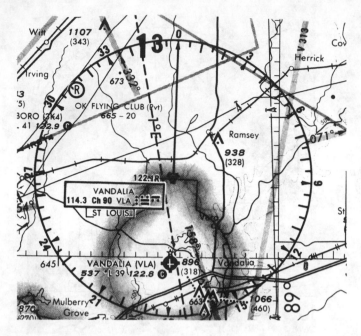

Fig. 4-5. *Vandalia is a case where you call the FSS on one frequency (122.1) and listen over the VOR (114.3).*

You: St. Louis Radio, Cherokee One Four Six One Tango on the ground, Vandalia, on 122.1, listening 114.3. [The last might not be necessary, because the FSS will probably know you're listening over the VOR since you're transmitting on 122.1 and are at Vandalia. It doesn't hurt to include the VOR frequency, though, just in case there's any possible confusion.]

Another alternative in this situation: It might be that the VOR is just out of radio range on the ground. Should that be the case, you'd have to take off and then make contact once airborne, with only a slight variation in the transmissions:

You: St. Louis Radio, Cherokee One Four Six One Tango on 122.1 Vandalia, listening 114.3.

FSS: Cherokee One Four Six One Tango, St. Louis Radio.

You: Radio, Cherokee Six One Tango was off Vandalia at fourteen twenty-five Zulu. Would you open my flight plan to Cincinnati at this time?

And so on.

Situation 4: No FSS on the Field; No RCO; No Adjacent VOR

In cases like this, ask the FSS briefer what frequency you should use after takeoff. It might be an RCO or the nearest VOR that has voice capabilities. If it is a fairly distant RCO, though, you'll probably have to get to 3,000 or 4,000 agl for reception, as the typical RCO range is rather limited. Whatever the case, your radio call follows the same format as in the Vandalia airborne example.

FILING A FLIGHT PLAN IN THE AIR

While it is possible to file a VFR or IFR flight plan while airborne, I
that you don't. Of course, if you're IFR qualified and run into unexpected IFR conui-
tions, that's one thing. Otherwise, refrain. Making contact with an FSS, going
through the acknowledgement, and reading off the information on the Flight Plan
form just takes time—time the FSS inflight specialist could be spending on other more
important issues.

In one case observed when visiting an AFSS, a specialist was occupied for over
10 minutes with an IFR pilot who was air-filing and asking all sorts of questions.
Meanwhile, the frequency was tied up and the specialist couldn't respond to other
calls. A little thoughtfulness, plus better preflight planning, should obviate the need
for any VFR filing in the air.

FLIGHT WATCH
(ENROUTE FLIGHT ADVISORY SERVICE, OR EFAS)

You're on a VFR flight from western Tennessee to Charlotte, North Carolina. As
you near Knoxville, you notice the afternoon buildup of cumulus activity ahead of you
over the Appalachian Mountains. Aware of the power of these summer-spawned thun-
derheads, you conclude that a little information about what's happening—or is
expected to happen—over your route is in order. So, Flight Watch comes into the picture. The only questions are, who do you call, and what do you say?

First, a bit of background. Not all AFSSs offer the Flight Watch service, with
only 44, as of the date of this writing, having been so designated. To determine those
that do have EFAS in the area in which you're flying, check the inside of the back
cover page of the appropriate *Airport/Facility Directory* (FIG. 4-6). For instance, in the
Southeast, there are 11 AFSSs, but only four—Jackson, Tennessee; Macon, Georgia;
Gainesville, Florida; and Miami, Florida—provide the service.

But what if you're a hundred or so miles from one of these AFSSs? How do you
make radio contact? The answer lies in the remote communications outlets that extend
outward from each AFSS. If you'll check FIG. 4-6 again, you'll note that an outlet in
the Knoxville (TYS) area connects you with the Macon AFSS. Or around Florence
(FLO), South Carolina, you'd be in contact with Gainesville (GNV), Florida.

Suppose, though, that you're about halfway between two of the outlets and aren't
sure which AFSS to call. In that case, dial in 122.0 anywhere in the country and
merely address the message to "Flight Watch," your aircraft N-number, followed by
the closest VOR to your position. For example: "Flight Watch, Cherokee One Four
Six One Tango at Hinch Mountain VOR." That identifies your location and the AFSS
servicing that area will respond.

To access Flight Watch, you should be between 5,000 feet agl and 17,500 feet
msl. Above 17,500, the AFSSs around the country provide "high altitude" EFAS on
frequencies other than 122.0. Also, the address should always be to the appropriate
Center, even though the AFSS will respond, followed by "Flight Watch," not ". . .

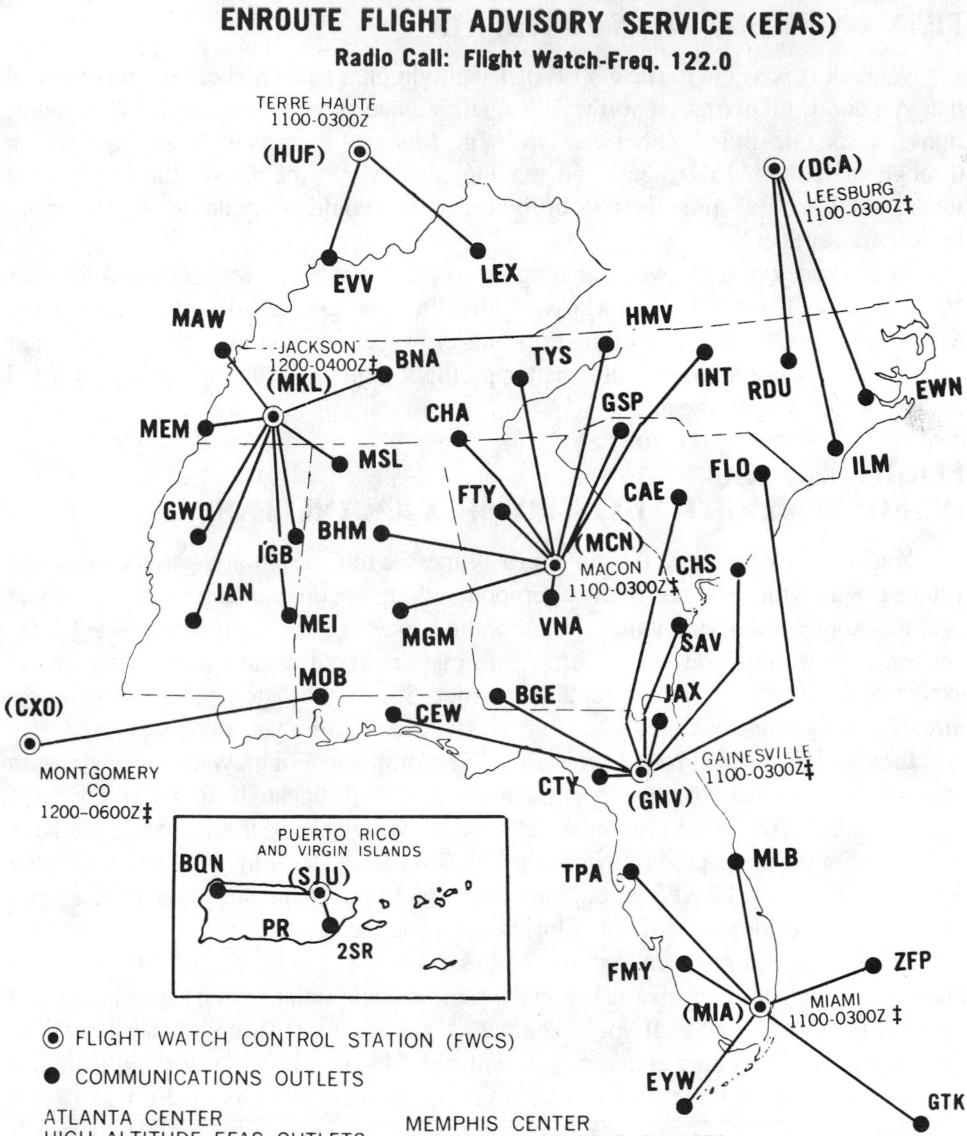

ENROUTE FLIGHT ADVISORY SERVICE (EFAS)
Radio Call: Flight Watch-Freq. 122.0

TERRE HAUTE
1100-0300Z
(HUF)

(DCA)
LEESBURG
1100-0300Z‡

EVV   LEX

MAW   HMV

JACKSON
1200-0400Z‡   BNA   TYS   INT   RDU   EWN

(MKL)

MEM   CHA   GSP

MSL   FLO   ILM

GWO   FTY   CAE

BHM   (MCN)   CHS

IGB   MACON
1100-0300Z‡

JAN   MEI   VNA   SAV

MGM   CHS

MOB   BGE   JAX

(CXO)   CEW

MONTGOMERY
CO
1200-0600Z‡   CTY   (GNV)   GAINESVILLE
1100-0300Z‡

MLB

PUERTO RICO
AND VIRGIN ISLANDS   TPA

BQN   (SJU)

PR   FMY   ZFP

2SR   (MIA)   MIAMI
1100-0300Z ‡

◉ FLIGHT WATCH CONTROL STATION (FWCS)   EYW   GTK

● COMMUNICATIONS OUTLETS

ATLANTA CENTER HIGH ALTITUDE EFAS OUTLETS		MEMPHIS CENTER HIGH ALTITUDE EFAS OUTLETS	
ATHENS	135.475	GRAHAM	133.675
CHATTANOOGA	135.475	GREENWOOD	133.675
MONTGOMERY	135.475		
TRI CITY	135.475	MIAMI CENTER HIGH ALTITUDE EFAS OUTLETS	
		AVON PARK	132.725
JACKSONVILLE CENTER HIGH ALTITUDE EFAS OUTLETS		MIAMI	132.725
CHARLESTON	134.175	WASHINGTON CENTER HIGH ALTITUDE EFAS OUTLETS	
GAINESVILLE	134.175	WILMINGTON	134.525
PANAMA CITY	134.175		
SAMPSON	134.175		

Fig. 4-6. *The nearest FSS Flight Watch to call can be determined by reference to the inside back cover of an* A/FD.

Radio." This is the one time when "Radio" is not used in conjunction with an FSS.

Back to your Nashville-Charlotte flight: You know that you're in Atlanta Center's area and, by reference to the *A/FD*, that the Macon AFSS provides EFAS. So, with 122.0 tuned in, you make the call:

You: Atlanta Flight Watch, Cherokee One Four Six One Tango.

FW: *Cherokee One Four Six One Tango, Atlanta (or Macon) Flight Watch.*

You: Flight Watch, Cherokee Six One Tango is over Oak Ridge at niner thousand five hundred, VFR to Charlotte. Request enroute weather advisories.

FW: *Cherokee Six One Tango, cumulus building northwest of Asheville and potential thunderstorms in the Snowbird VOR area. Asheville is clear at this time, but expect widely scattered thunderstorms by sixteen hundred local. Winds two two zero at niner thousand. Ceilings and visibility unlimited Asheville to Charlotte. Asheville altimeter three zero two five, Macon (or Atlanta) Flight Watch.*

You: Roger, Flight Watch. Thank you. Cherokee Six One Tango.

A point to remember about Flight Watch is that this is strictly a weather-related service—nothing more. Since that is its only role, never use it for routine position-reporting, filing or amending a flight plan, an emergency, or any other purpose. Those contacts should be made on 122.2 or one of the published FSS frequencies, while the emergency frequency is universally 121.5.

OTHER INFLIGHT WEATHER ADVISORIES

Flight Watch isn't the only source of weather information while in flight. For example, if you'll check FIG. 4-7, you'll notice the small square in the lower right corner of the Knoxville VOR identification box. This, as mentioned earlier, indicates the continuous recorded transmission of HIWAS (Hazardous Inflight Weather Advisory Service). The formerly broadcast TWEB has been, or is being, phased out.

Fig. 4-7. *The small square in the VOR identification rectangle indicates the availability of HIWAS.*

In essence, the HIWAS transcriptions contain summaries of hazardous weather within a given Air Route Traffic Control Center's area of responsibility, based on the various weather-reporting/forecasting sources. Also included are isolated thunderstorms and limited areas of low ceilings and visibilities that would not normally warrant a special weather advisory broadcast. The message concludes with: "Contact

Flight Watch or Flight Service for additional details. Pilot weather reports are requested." When there are no pertinent weather advisories, a statement to that effect is broadcast hourly.

While it is beyond the scope of this book to get into meteorological discussions or to detail all of the weather-reporting vehicles, a few of the types of inflight advisories available to the pilot must be mentioned. The following are either summarized, as pertinent, in the HIWAS alert message or the specifics of each can be obtained by contacting Flight Watch or an FSS:

- AIRMETs reflect conditions that present a potential hazard to light planes and to those lacking instrumentation or equipment (as deicers or anti-icers). In effect, an AIRMET is an amendment to the area forecast.

- SIGMETs report significant meteorological developments that could be particularly hazardous to light aircraft and potentially hazardous to all aircraft. They also are included in the area forecast.

All Flight Service Stations within 150 miles of the weather area broadcast SIGMETs upon issuance and at 15 and 45 minutes past the hour during the first hour. Thereafter, an alert notice is broadcast at 15 and 45 minutes past the hour (for the duration of the advisory), stating that "SIGMET (name and number) is current."

If you're monitoring an FSS or a VOR and pick up only the alert notice (not having heard the initial and complete SIGMET or AIRMET), you would be wise to call either the FSS or Flight Watch for further information:

You: De Ridder Radio, Cherokee One Four Six One Tango on one two two point two.

FSS: Cherokee One Four Six One Tango, De Ridder Radio.

You: De Ridder, Cherokee Six One Tango. Would you read me SIGMET Delta Four?

FSS: Cherokee Six One Tango, SIGMET Delta Four reads as follows: Flight precaution eastern Louisiana, moderate to severe turbulence in clouds seven thousand to fifteen thousand feet msl. Conditions expected to continue until zero three hundred Zulu.

You: Roger, De Ridder. Cherokee Six One Tango.

SIGMETs report severe to extreme turbulence, severe icing, and dust or sand-storms that reduce visibility below 3 miles. In effect, they report all of the more hazardous conditions *except* thunderstorms, tornadoes, and hail.

Convective SIGMETs focus on these latter phenomena, and, as with AIRMETS and SIGMETS, are broadcast on the voice facilities of Flight Service Stations, as well as being available for the pilot's preflight review. Beginning daily at 0000Z, convective SIGMETs are numbered consecutively from 01 to 99, with each valid for one hour.

PIREPs (Pilot reports) are perhaps the most significant and accurate sources of current flight conditions. Who knows better than the pilot who is there now and can report what is actually happening? Light turbulence that was forecast turns out to be moderate or severe. Unpredicted icing is experienced. Winds that were supposed to be from 240 degrees at 15 knots suddenly become brutal headwinds or helpful tailwinds at 45 knots. A flock of waterfowl finds its way into the traffic pattern. A sudden wind shear is encountered on takeoff or landing. All of these unexpected events should be immediately reported to the tower, Flight Service, or unicom (depending on the facility at the particular airport).

As a general rule, submit a PIREP whenever conditions occur that were not forecast or that are actually or potentially hazardous to flight—weather-related or otherwise. Current PIREPs help forecasters, briefers, controllers, and pilots alike. (Flight Watch even appreciates reports of favorable conditions.)

What should be reported, keeping in mind the general rule stated above? Basically, unanticipated icing, clear air turbulence, wind shears, thunderstorms, ceiling or visibility changes (from those forecast), significant wind direction or velocity changes, in-flight or runway obstacles, precipitation, and anything else you encounter that was unexpected and could present a danger to other pilots.

To whom should you make the report? Depending on where you are and with whom you are in communication, the report should go to the nearest Flight Service Station, the Air Route Traffic Control Center (if you're using this facility), Flight Watch, the Control Tower, or, if at an uncontrolled airport, the local unicom. These are the sources that will make use of your report and convey it to others orally and (except for unicom operators) over the weather communications network.

An example of a PIREP communicated in flight:

You: Wichita Radio, Cherokee One Four Six One Tango on one two two point four.

FSS: *Cherokee One Four Six One Tango, Wichita Radio.*

You: Wichita, Cherokee Six One Tango PIREP. Cherokee one eighty over Abilene at two two three zero Zulu, encountering continuous moderate chop between Topeka and present location. Clear of clouds at eight thousand five hundred.

FSS: *Cherokee Six One Tango, Roger. Keep us advised if conditions continue or worsen.*

You: Roger, will do. Cherokee Six One Tango.

In the PIREP, include:

- Type of aircraft
- Location
- Time (UTC)
- Conditions you are reporting
- Whether in or out of clouds

- Altitude
- Duration of conditions you are reporting

If the PIREP concerns icing or turbulence, use the approved definitions as detailed in the *Airman's Information Manual*. What to you might seem to be severe turbulence could, by definition, be moderate or perhaps even light. Just be sure that you and the person on the ground are speaking the same language. Misuse of official terms because of unfamiliarity with them can result in inaccurate information to other pilots. They are either lulled into a sense of false security (and we know the dangers of that), or they are warned of conditions that don't actually exist, which could cause flight deviations, increased time enroute, and increased fuel consumption.

So you should know what you're talking about when you submit a PIREP. But even if you're not certain about the exact conditions you're encountering, according to official definition, a slightly inaccurate PIREP, when unexpected conditions arise, is better than no PIREP at all. Others will appreciate your concern for their well-being.

HIWAS (Hazardous In-Flight Weather Advisory Service), a new TWEB-like continuous broadcast service, is becoming increasingly available over many navaids to disseminate important weather advisories, such as SIGMETs and AIRMETs.

EXTENDING THE FLIGHT PLAN

You're over central Kansas, enroute to Kansas City on a 3-hour flight plan. The winds, however, aren't holding up to the forecast. Either the tailwind is less or the headwinds are greater. Whatever the reason, you're not going to make the 3-hour estimate. It looks like the enroute time will be closer to $3^1/2$ hours, if not a little longer.

Keeping in mind that the FSS starts asking questions if you haven't closed out your flight plan within 30 minutes of your estimated arrival, you decide to extend the flight plan, based on the new ETA. If you're in the vicinity of Hays, the call would go like this:

You: Wichita Radio, Cherokee One Four Six One Tango, on one one zero point four, Hays.

FSS: *Cherokee One Four Six One Tango, Wichita Radio.*

You: Wichita, Cherokee Six One Tango over Hays at seven thousand five hundred on VFR flight plan to Kansas City Downtown, with a one five zero zero local ETA. Would like to extend the ETA to one five three zero.

FSS: *Cherokee Six One Tango, Roger. Will extend your VFR flight plan to Kansas City to one five three zero local.*

You: One Five Three Zero. Roger, thank you. Cherokee Six One Tango.

AMENDING THE FLIGHT PLAN IN FLIGHT

The flight is from Memphis to Kansas City, a distance of 410 statute miles. With forecast winds of about 20 knots from 270 to 290, you plan to make a pit stop at

Springfield, Missouri. Along the way, however, you find the winds are much less than anticipated and that you have plenty of fuel to reach Kansas City nonstop. After passing the Dogwood VOR, 35 miles southeast of Springfield, you tune to the Springfield VOR and continue on Victor 159 toward the station.

In checking the sectional, you note the solid square in the lower right corner of the VOR box, indicating that the station transmits HIWAS. With this service available to you, you monitor the broadcast on the VOR frequency and find that the conditions into Kansas City justify the elimination of the Springfield stop. So you call the Columbia FSS:

You: Columbia Radio, Cherokee One Four Six One Tango on one two two point five five, Springfield.

FSS: *Cherokee One Four Six One Tango, Columbia Radio.*

You: Columbia, Cherokee One Four Six One Tango is two zero miles southeast on Victor One Five Niner at eight thousand five hundred on VFR flight plan Memphis to Springfield. We have two point five hours of fuel remaining and would like to amend our flight plan and proceed direct to Kansas City Downtown, with an ETA of one four four five local.

FSS: *Cherokee Six One Tango. Understand you have two point five hours of fuel remaining. We will amend your flight plan to show you direct to Kansas City Downtown with a one four four five ETA.*

You: Roger. Thank you for your help. Cherokee Six One Tango.

CHECKING RESTRICTED AREAS OR MILITARY OPERATIONS AREAS (MOAs)

The country is full of areas that are designated as *Prohibited, Restricted, Warning,* or *Alert*. And there are also the *Military Operations Areas*. Suffice it to say that flight into a Prohibited Area is *prohibited*; these are sufficiently few in number and limited in size that they are generally easy to circumnavigate. Many Restricted Areas are also relatively easy to avoid. If, however, flight through such an area is essential, authorization from the controlling agency is required. Entering an active Restricted Area without authority could expose you to all sorts of military hazards, from aerial gunnery to guided missiles. By checking the Special Use Airspace table on the sectional, you'll find that most Restricted Areas have regular hours of use, but some can be activated by NOTAM on an irregular basis. So you'll need to check with an FSS or, in flight, with the controlling agency listed in the table. In most cases, the controlling agency is the Air Route Traffic Control Center.

Warning Areas are located offshore beyond the 3-mile limit, but Alert Areas can be found throughout the country. Normally, the latter define the geography over which there is a high volume of flight training or unusual aerial activity. Transiting these areas is not prohibited, nor is special authorization required. The pilot, however, is urged to exercise extreme vigilance and caution because of the activity. The responsibility for collision avoidance is his.

Military Operations Areas present the largest geographical obstacle, and to circumvent one might take you miles out of your way. Rather than consume fuel that could otherwise be conserved, the safety of passage through or into an MOA can be determined by contacting any FSS within 100 nautical miles of the area. Perhaps you can do this on the ground during the preflight briefing. Otherwise, call the nearest FSS before entering the MOA.

For example: You're making a trip from Memphis to Jacksonville, Florida. The most direct routing is to Birmingham, Tuskegee, Albany (Georgia), and on to Jacksonville. Just east of Albany, however, lies the extensive Moody MOA—Moody 1, 2A, and 2B. To avoid the MOA would add at least 55 statute miles to the flight—not a major detour, but a needless one in time and fuel, if it can be avoided.

The answer, then, is to contact Macon AFSS to determine the current activity in the area and the extent to which passage through it is feasible. The commonly used word to denote the existence of military operations is "hot."

You: Macon Radio, Cherokee One Four Six One Tango on one two two point two over Albany VOR.

FSS: *Cherokee One Four Six One Tango, Macon Radio.*

You: Macon, Cherokee One Four Six One Tango is ten miles northwest on Victor One Five Niner at seven thousand five hundred enroute Jacksonville on VFR flight plan. Can you advise if the Moody MOAs are hot?

FSS: *Cherokee Six One Tango, affirmative. Moody 1A and 2A are both hot at this time. Contact Jacksonville Center for advisories, frequency one three two point five five.*

You: Roger, will do. Cherokee Six One Tango.

Obviously, what Jacksonville Center tells you will dictate your actions—to go through the MOA, or take the longer way around. You're not prohibited from entering the MOA, but if there's enough high-speed activity buzzing around in there, good judgment might tell you to follow the airways and add another half hour or so to your journey.

CLOSING OUT THE FLIGHT PLAN

One way to close out a flight plan is by telephone at your destination. The other way is via radio after you're on the ground and parked at the ramp. There can be some exceptions when an inflight close-out is in order, but the best approach is to wait until you've landed and the flight is completed.

Whichever method you choose, just be sure you do close it out! Failure to do so sets a lot of wheels in motion to track you down, and if it turns out to be unnecessary, it does not sit very well with the powers that be.

If the FSS or an RCO is on the field and you select the radio, the call is simply this:

You: Jackson Radio, Cherokee One Four Six One Tango on one two two point two; Memphis.

FSS: *Cherokee One Four Six One Tango, Jackson Radio.*

You: Cherokee One Four Six One Tango is on the ground at Memphis International. Please close out my VFR flight plan from Springfield, Missouri, at this time.

FSS: *Cherokee Six One Tango, Roger. Closing out your flight plan at three five.*

You: Understand we're closed at three five. Thank you. Cherokee Six One Tango.

OBTAINING SPECIAL VFR:
Flight service remoted, no tower

You want to depart from an airport within a Control Zone. (See Chapter 8 for a discussion of control zones.) No tower is on the field, the FSS is physically located elsewhere, and the weather is less than 1,000 feet and 3 miles visibility. Once clear of the Control Zone, you know that you can maintain legal VFR limits of visibility and cloud separation. How do you obtain a clearance under these conditions?

Let's take a specific example. Gage, Oklahoma's Gage-Shattuck Airport, has only unicom—no tower, no Flight Service Station. There is, however, a Remote Communications Outlet through which you can contact the McAlester FSS more than 200 miles to the southeast. The remoted facility is identified on the sectional (FIG. 4-8).

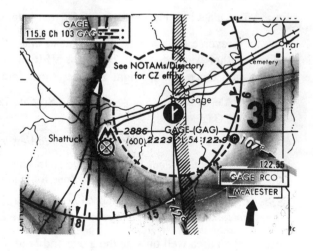

Fig. 4-8. *The Gage airport has a Control Zone, the FSS is in McAlester, Oklahoma, but an RCO is located on the airport.*

The weather being what it is, you can't legally take off in the Control Zone without clearance through Flight Service. Hence, this call:

You: McAlester Radio, Cherokee One Four Six One Tango, one two two point five five, Gage.

FSS: *Cherokee One Four Six One Tango, McAlester Radio.*

You: McAlester, Cherokee One Four Six One Tango requests Special VFR out of the Gage Control Zone, west departure.

FSS: *Cherokee Six One Tango, stand by. (FSS now contacts the appropriate control agency.)*

FSS: *Cherokee Six One Tango, ATC clears Cherokee One Four Six One Tango to exit the Gage Control Zone to the west. Maintain Special VFR conditions at or below, three thousand feet while in the Control Zone. Report leaving the Control Zone.*

You: Roger, understand Cherokee Six One Tango is cleared out of the Gage Control Zone to the west, to maintain Special VFR at or below three thousand in the Zone. Will report leaving the Zone.

FSS: *Cherokee Six One Tango, readback is correct.*

You: Roger. Cherokee Six One Tango.

In a nonradar area, only one aircraft at a time is allowed in the Control Zone when conditions are below normal VFR limits. Thus, for safety first and courtesy to others second, it is imperative to advise Flight Service when you are clear of the Zone.

You: McAlester Radio, Cherokee One Four Six One Tango is clear of the Gage Control Zone to the west.

FSS: *Cherokee Six One Tango, Roger. Gage altimeter two niner zero eight.*

You: Two niner zero eight. Thank you for your help. Cherokee Six One Tango.

Now let's reverse the process. You want to enter the Control Zone and land at Gage. The weather was forecast to be below VFR limits, and a call to Gage unicom verifies that fact. Once again, Flight Service comes into the picture:

You: McAlester Radio, Cherokee One Four Six One Tango one two two point five five, Gage.

FSS: *Cherokee One Four Six One Tango, McAlester Radio.*

You: Cherokee One Four Six One Tango is 15 miles west of Gage and requests Special VFR to enter the Control Zone for landing at Gage.

FSS: *Cherokee Six One Tango, we have one other aircraft in the Control Zone for landing. Stand by and remain clear of the Zone until further advised.*

Now, find an area well outside the Zone and hold until the FSS authorizes you to proceed:

FSS: *Cherokee One Four Six One Tango, McAlester Radio.*

You: McAlester Radio, Cherokee One Four Six One Tango, go ahead.

FSS: *Cherokee Six One Tango, ATC clears Cherokee One Four Six One Tango to enter the Gage Control Zone. Maintain Special VFR conditions at or below, three thousand feet while in the Control Zone. Report when down and clear at Gage.*

You: Understand, Cherokee Six One Tango cleared to enter the Gage Control Zone at or below three thousand on Special VFR. Will report on the ground at Gage.

FSS: Cherokee Six One Tango, readback correct.

You: Roger. Cherokee Six One Tango.

You're on the ground and clear of the active at Gage. Now call Flight Service over the RCO as quickly as possible so that another aircraft can be cleared into or out of the Zone.

You: McAlester Radio, Cherokee One Four Six One Tango is down and clear at Gage.

FSS: Understand Cherokee Six One Tango is down and clear at Gage.

You: That is affirmative—and thank you for your help. Cherokee Six One Tango.

OBTAINING SPECIAL VFR:
Flight service on the airport, no tower

Take the example of Bowling Green, Kentucky. This is a unicom, no-tower airport, but one of the permanent XFSSs is located on field (FIG. 4-9). Before departing Bowling Green, you check the weather with Flight Service, file a flight plan, if necessary, and, since the local conditions are below the VFR minimums of a 1,000 foot and 3 miles visibility, you advise the specialist that you'll be requesting a Special VFR. Even though the FSS requested the SVFR while you were in its facility, it's unlikely that the controlling ATG would approve it at that time. The reasons: At the time you were ready to go, the minimums might drop below SVFR limits; your estimated departure might be delayed; or other aircraft could be entering or leaving the Control Zone.

Once you are ready, though, the radio contacts are the same as those illustrated in the Gage example. Whether an FSS is on the field or remoted makes no difference. The procedures for departing and entering the Zone in less-than-VFR but within SVFR conditions don't vary.

OBTAINING AIRPORT ADVISORIES:
Flight service on the airport, no tower

When an operating Flight Service Station is located on the airport and there's no operating tower, as at Bowling Green, the FSS provides the same airport advisory that unicom offers at fields with no tower or FSS, but it's more detailed and more accurate.

If you're landing at an airport with an FSS and want an airport advisory, what frequency do you use? The answer is the same at all FSS airports: 123.6. This is the

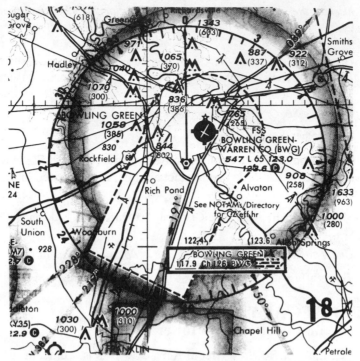

Fig. 4-9. An FSS is located on the field at Bowling Green, and can be reached on 123.6.

standard frequency for the AAS, or Airport Advisory Service. It is not, however, the frequency for filing or closing out flight plans or for obtaining enroute weather information. Going into a field which has both Flight Service and unicom, use the latter for phone messages, taxis, and the like, and the former for the advisory service. Accordingly, and on 123.6, the communication goes like this:

> **You:** Bowling Green Radio, Cherokee One Four Six One Tango on one two three point six.

> **FSS:** *Cherokee One Four Six One Tango, Bowling Green Radio.*

> **You:** Bowling Green, Cherokee Six One Tango is ten miles west of Bowling Green VOR at three thousand. Request airport advisory.

> **FSS:** *Cherokee Six One Tango, favored runway is Two One, wind one niner zero at one two, altimeter two niner eight seven. No reported traffic.*

> **You:** Roger. Thank you. Cherokee Six One Tango.

KEEPING LOCAL TRAFFIC INFORMED

Remember that Flight Service is in no way a traffic control agency at Bowling Green or anywhere else. So once you enter the pattern, stay on 123.6 but address your communiques to "Bowling Green Traffic," not "Bowling Green Radio."

From this point on, the calls are the same as at a multicom or unicom field—entering the pattern on downwind, turning base, turning final, and on the ground and clear of the runway.

As with unicom, don't expect any response from the FSS when you make these transmissions. I've noticed on several occasions, however, particularly at the smaller or less busy airports, that the specialist does acknowledge even routine position reports. This despite the pilot's use of "Traffic" rather than "Radio." So, if the FSS does respond to or acknowledge blind transmissions, don't be surprised. On the other hand, don't be surprised if silence reigns. That's the way it should be.

Remember, as with unicom, that there's no requirement for you to contact Flight Service for an advisory. Not doing so is a little foolish, though. Fulfill your responsibility, but keep your eyes open and your head turning to spot any possible Silent Sams.

CLOSING OUT A VFR FLIGHT PLAN:
Flight service on the airport, no tower

At the Bowling Greens of the nation, when you land and want to cancel your flight plan, you can do so by radio, if that's your choice. But just be sure to switch from 123.6 to another published frequency—in the case of Bowling Green, 122.4.

You: Bowling Green Radio, Cherokee One Four Six One Tango on one two two point four, Bowling Green.

FSS: *Cherokee One Four Six One Tango, Bowling Green Radio.*

You: Cherokee One Four Six One Tango is on the ground at Bowling Green. Would you close my flight plan from Lexington at this time?

FSS: *Cherokee Six One Tango, Roger. Will close out your flight plan at two seven.*

You: Roger. Thank you. Cherokee Six One Tango.

TAXIING OUT AND BACK-TAXIING:
Flight service on the airport, no tower

As we said in the multicom and unicom chapters, if you're going to back-taxi, complete the entire pretakeoff check before venturing onto the runway—unless an area exists at the end of the runway for that purpose. It's simply discourteous to others and disruptive to traffic to park on the approach end while you go through the checklist. More than one landing pilot has been forced to go around because some self-centered idiot was sitting there on the runway, oblivious to the existence of all others. If there is no runup area, make the check on the ramp or taxiway, back-taxi with some speed and power, do a 180, and get going.

But to begin at the beginning: you're on the ramp, engine started, and you want an airport advisory. Tune to 123.6 and call Flight Service:

You: Bowling Green Radio, Cherokee One Four Six One Tango on the ramp, ready to taxi for takeoff. VFR northbound. Request airport advisory.

> **FSS:** *Cherokee Six One Tango, favored runway is Two Three, wind one niner zero at four. Altimeter three zero zero zero. Mooney reported ten west for landing.*

> **You:** Roger, will back-taxi on Two Three after runup. Cherokee Six One Tango.

You've completed the pretakeoff check and are ready to taxi:

> **You:** Bowling Green Traffic, Cherokee Six One Tango back-taxiing Runway Two Three, Bowling Green.

You're ready to go:

> **You:** Bowling Green Traffic, Cherokee Six One Tango taking Two Three, northwest departure, Bowling Green.

CTAFs: Part/full-time FSS, part/full-time tower

Budget-cutting in recent years has brought about a profusion of part-time towers and part-time FSSs. As a general rule, when a tower exists on a field, even if it is closed, use the tower frequency for traffic advisories. If there is no tower but an FSS exists on the field, use 123.6 for self-announce advisories even if the FSS is closed. TABLE 4-1 summarizes the various possibilities, but remember to check the *A/FD* for exceptions.

Table 4-1. Frequencies for traffic advisories

Tower status	FSS status	Frequency
Open tower on field		Tower frequency
No tower on field	Open FSS on field	AAS on 123.6
No tower on field	Closed FSS on field	Self-announce on 123.6; advisories from unicom (if available)
No tower on field	No FSS on field	Unicom (if available) or multicom (check A/FD)
Closed tower on field	Open FSS on field	AAS on tower frequency
Closed tower on field	Closed (or no) FSS on field	Self-announce on tower frequency

TABLE 4-2 helps keep things straight when there is no tower, but a Flight Service Station as well as unicom are on the field.

DIRECTION FINDING (DF) ORIENTATION

You're lost or unsure of your position. As a rule, if you have even one navcom on board, you should be able to reorient yourself with little difficulty. It's just a matter of

Table 4-2. FSS and unicom radio calls

Purpose of call	Frequency	Radio call
For airport advisories (when FSS is open)	123.6	"_____ Radio"
Traffic position reports (at all times)	123.6	"_____ Traffic"
Flight plan filing/closing	Published FSS frequency or 122.2	"_____ Radio"
To order taxis, relay messages, request mechanics, etc. (and airport advisories when FSS is closed)	Published unicom frequency (usually 122.7, 122.8, or 123.0)	"_____ Unicom"

tuning to one VOR, centering the needle, drawing the radial from that station on your sectional, and then doing the same thing with a second VOR. Your position is where the lines intersect. Sometimes, however, you can be in an area or at an altitude where one or more VORs are out of receiving range, or, indeed, the navcoms could go kaput. Now you might need help. Here is where a Flight Service Station can provide emergency assistance service, help you get oriented, and, if necessary, provide you with a DF steer to the nearest airport.

Briefly summarized, depending on the ground equipment available, the FSS, in coordination with the pilot, offers five types of DF orientation:

1. Almost immediate location determination with the FSS using two geographically separated DF receivers. One receiver is usually in the FSS facility, the other at a remote location.

2. Orientation with one DF receiver and the pilot tuning to a VOR, when that VOR is within range of the aircraft's reception, to obtain the radial from the VOR, and thus a "cross-fix" identifying the aircraft's position.

3. DF-T/D (time and distance), when the FSS uses only one DF and no other means of obtaining a cross-fix is available.

4. DF-ADF, using one DF receiver, with the pilot tuning to a nondirectional beacon (NDB) within the aircraft's reception range—if the aircraft is equipped with automatic direction finding (ADF) equipment. This produces a cross-fix similar to the DF-VOR orientation in number 2 above.

5. VOR-to-VOR, when no DF equipment is available (or is beyond the aircraft's range), the aircraft is within range of two VORs, and has at least one operating VOR receiver on board. This type of orientation the pilot can do by himself, if

he has a current sectional on board, but inexperience, increasing anxiety, or other factors might require the help of the FSS specialist.

Rather than trying to illustrate what would take place in each of the five orientation possibilities, let's focus just on the first, where there is a VOR on board and the FSS uses two DF receivers—one at the station and one remote. The situation, then, follows:

You're heading west in western Iowa, where the land is flat and easy-to-spot landmarks are infrequent. You know that you're somewhere southwest of Fort Dodge, but your one VOR head is proving unreliable, a study of the terrain reveals nothing identifiable, and your fuel is rapidly diminishing. Time is critical, so you call the Fort Dodge (FOD) Automated Flight Service Station for emergency assistance.

First, what is the general sequence of information the AFSS specialist will request, and what instructions will he likely issue? There might well be variances, but the following is in line with current AFSS procedures:

1. The *minimum* information the AFSS will want is aircraft identification, aircraft type, if transponder-equipped, pilot intentions, and nature of the emergency. Additionally, the specialist might ask the weather at your altitude, the hours of fuel remaining, and possibly (or probably) to squawk the "mayday" 7700 transponder code. If the last is the case, the specialist will then coordinate the orientation process with the appropriate Center or airport Approach Control.

2. Other likely instructions would tell you to maintain VFR flight, advise of any necessary altitude changes, and advise if instructions would cause you to violate VFR regulations. The specialist will also give you the current altimeter setting, ask you to set the directional gyro heading indicator to the magnetic compass heading, and then request you to report current heading and altitude.

3. Instructions will be included for keying the microphone by depressing the mike button for 10 seconds, saying nothing during that period, and concluding the transmission with your aircraft identification.

4. And there will be instructions during the actual orientation process. If the FSS is using two DF receivers, your position will be determined very rapidly. More extensive directions are probable if the FSS is using only one DF and wants you to tune to a VOR for a cross-fix. The same applies when the orientation involves one DF and an NDB, a DF/time-and-distance procedure, or the VOR-to-VOR method.

5. Finally, if necessary, the FSS will give you headings, or a DF steer, to the nearest airport or the airport of your choice, depending upon the situation.

You: Fort Dodge Radio, Cherokee One Four Six One Tango on one two two point two.

FOD: *Cherokee One Four Six One Tango, Fort Dodge Radio.*

You: Fort Dodge, Cherokee Six One Tango is somewhere southwest of you at six thousand five hundred. VOR unreliable, and have three zero minutes of fuel remaining. Request a DF steer to the nearest airport.

FOD: *Roger, Six One Tango. Are you transponder-equipped?*

You: Affirmative, Fort Dodge.

FOD: *Six One Tango, squawk seven seven zero zero, for one minute, then seven six zero zero.*

You: Roger, seven seven zero zero. Six One Tango.

FOD: *Six One Tango, what are the weather conditions at your altitude?*

You: Scattered cumulus, Fort Dodge.

FOD: *Roger, Six One Tango. Remain VFR at all times, and advise if you have to change heading or altitude to remain VFR. Fort Dodge altimeter is three zero zero five.*

You: Roger, three zero zero five.

FOD: *Six One Tango, maintain straight and level flight and reset your directional gyro to agree with your magnetic compass. After you have done this, say your heading and altitude.*

You: Fort Dodge, Six One Tango heading is two seven zero degrees, level at six thousand five hundred.

FOD: *Roger, Six One Tango. Now depress your mike button for ten seconds, saying nothing, and then follow with your aircraft identification.*

You: Roger. [Depress the mike button ten seconds.] Cherokee One Four Six One Tango.

Note: During the 10 seconds, the AFSS specialist is watching the strobes on the two DF receivers, as the strobes search out your aircraft based on your radio transmission. Where the strobes intersect identifies your position. The specialist then plots that position on a sectional chart.

FOD: *Cherokee Six One Tango, your position is one five miles northeast of the Denison Airport. What are your intentions?*

You: We need to land Denison as soon as possible for fuel. Six One Tango.

FOD: *Roger, Six One Tango. For a heading to the Denison Airport, turn left to two two zero degrees. Report when on that heading.*

You: Roger, left to two two zero.

You: Fort Dodge, Six One Tango on the two two zero heading.

FOD: *Roger, Six One Tango. Do you see any prominent landmarks?*

You: Affirmative. I see a water tower and a town in the distance at twelve o'clock.

FOD: *Roger, Six One Tango. That should be Denison. Are you familiar with the airport?*

You: Negative. Please advise.

FOD: *Roger, Six One Tango, the Denison Airport CTAF is two miles southwest of the town, elevation one thousand two hundred seventy three feet. The one runway is one two and three zero. Report when the airport is in sight.*

You: Roger, will do.

You: Fort Dodge, Six One Tango has the airport.

FOD: *Roger, Six One Tango. The Denison unicom CTAF is one two two point eight and Denison altimeter is three zero zero zero.*

You: One two two point eight and three zero zero zero.

FOD: *Roger, Six One Tango, do you require any further assistance?*

You: Negative, Fort Dodge. Six One Tango is in good shape now. Thank you for your help.

FOD: *Roger, Six One Tango. DF orientation service is terminated. Would you contact us by phone when you're on the ground?*

You: Will do.

The reason for this last request is that the AFSS has to fill out a Flight Assist Report for statistical purposes, and to do so, the specialist needs to ask some questions that the pilot might prefer not to answer over the air. There is no negative connotation or threat of disciplinary action behind the request. It is strictly for statistics, and thus no pilot should be concerned about making the call and responding to the questions.

THE FSS AND POSITION-REPORTING

Let's say that you're on a nonstop cross-country from Alpha City to Charlie City and the route takes you over or near Bravo Town. It's a 400-mile jaunt which, based on your briefing, should take about four hours. So you file a flight plan with a 1200 hour departure and 1600 hour ETA. You open the flight plan by radio with FSS Alpha at noon, and Alpha then sends the pertinent operational data, including your 1600 hour ETA, to the FSS responsible for the area in which Charlie is located. Once you're under way, is there any benefit to reporting your position to an FSS as you move down the road? Potentially, yes.

Suppose you're between Bravo and Charlie. You haven't contacted any facility since leaving Alpha, but suddenly you encounter a major mechanical emergency. There's no time to think about radio calls or squawking the emergency 7700 transponder code. All that matters is getting that airplane on the ground now, hopefully in one piece. So down you go, ending up in some desolate plot of landscape.

An hour or so later, the 1600 ETA comes and goes with no word from you. At 1630, your computerized flight plan alerts the Charlie FSS that the flight plan has not been closed. A Charlie specialist then gets on the phone to see if you actually departed Alpha or if you had indeed landed at Charlie but had failed to close out the flight plan.

These efforts proving fruitless, the calls and messages start flowing to determine if any facility along your route has heard from you. If not, the Search-and-Rescue (SAR) unit serving that area is alerted to a possible downed aircraft. Maybe the Air Force gets into the act. Calls are made to your family or business associates, hoping for helpful information. With nothing forthcoming, the physical search begins.

Since no one knows where you are between Alpha and Charlie, the search has to cover the whole 400-mile stretch of territory. Meanwhile, you might be lying injured in a wrapped-up piece of fuselage that is no more than a speck of metal to those in the airborne SAR mission.

Time being critical in such an instance, just one enroute position report could be a life-saver. Had you contacted the FSS serving the Bravo area, which, let's say, is half-way between Alpha and Charlie, the search effort would have gone on as indicated, but it would have been only between Bravo and Charlie rather than over the whole 400-mile stretch. That one radio call, then, theoretically doubled your chances of an early rescue. And the call is no more complicated than this:

You: Bravo Radio, Cherokee One Four Six One Tango on one twenty two point two at Bravo. Position report.

FSS: Cherokee One Four Six One Tango, Bravo Radio. Go ahead.

You: Bravo, Cherokee Six One Tango is ten north of Bravo Town at one four two five local, VFR to Charlie City. Six One Tango.

FSS: Roger, Cherokee Six One Tango. Over Bravo at one four two five local. Bravo altimeter two niner four five.

You: Two niner four five. Roger, Bravo. Thank you. Six One Tango.

The Bravo FSS now has your position and time on tape, in the event either is needed in the future.

Disregarding for the moment the safety value of position reports, the importance of closing out a flight plan perhaps becomes obvious. The same process summarized above is initiated 30 minutes after your ETA has expired and no FSS has heard from you. Obviously, it can be a costly proposition when you fail to make that last FSS phone or radio call, no matter where you may have eventually landed. Always close out your flight plan with the nearest FSS!

I'll talk more about emergency actions in chapters 7 and 13, as well as the advantages of being in contact with the Air Route Traffic Control Centers and how those facilities can help you when things go wrong. Meanwhile, your friendly FSS represents, in all respects, a pretty solid insurance policy. I hope you'll agree that it's a policy worth buying when you venture forth across the airways.

5
Automatic Terminal Information Service (ATIS)

Before continuing, I'd like to indicate that this book has a certain sequential flow or logic. Essentially, it attempts to take you from the simplest of all of radio communications at multicom and unicom airports up through the use of Approach/Departure Control and Center on a cross-country flight.

It's not always feasible, however, to maintain that exact sequence, as chapter 4 on Flight Service Stations, illustrates. In describing the various FSS services, it seemed practical to include examples of typical inflight and route communication exchanges that can take place between a pilot and an FSS specialist. In that sense, the intended sequence was violated. On the other hand, for cross-country flights beyond the local airport vicinity, the FSS is typically the first contact a pilot has with the National Airspace System, because it's here where the weather is determined and the flight plan is filed.

From now on, however, and to maintain sequence, the other ATC services will be reviewed in the order that you would or could encounter them on either a local flight or a cross-country excursion. Beginning with this chapter, these facilities and services include ATIS, Ground Control, the tower, Approach and Departure Control, concluding with the Air Route Traffic Control Center. Finally, as sort of a review, I'll wrap everything up with a simulated cross-country that illustrates all of the radio contacts from engine start-up to final shut-down.

WHAT IS ATIS?

The Automatic Terminal Information Service, with its neat little acronym of ATIS, is hardly a stranger to most pilots—especially those acquainted with controlled airports. As a source of important information, it is a vital communications tool for both the departing and arriving airman.

For those not very familiar with it, ATIS is a continuous recorded summary, not more than one hour old, of local weather, winds, and the runway(s) and instrument approach(es) currently in use. Its purpose is to provide the pilot with the basic airport data on a frequency that does not interfere with live radio communications. The controllers then can concentrate on controlling traffic without having their attention diverted by constantly repeating that basic airport information.

Recorded in—and transmitted from—most but not all the towers, ATIS information is normally updated every hour, and is identified by the phonetic alphabet: Information Alpha, Bravo, Charlie, etc. Should conditions change in any material way before the next scheduled update, a revision is issued accordingly and given the next phonetic designation.

DETERMINING THE APPROPRIATE FREQUENCY

If ATIS is available at a given airport, its frequency is indicated on the sectional and in the *Airport/Facility Directory*.

To illustrate from the sectional, take the central Nebraska Regional Airport in Grand Island as an example (FIG. 5-1). In this case, the ATIS is clearly identified as 120.65. If no frequency is stated, the service is not provided, and winds, altimeter,

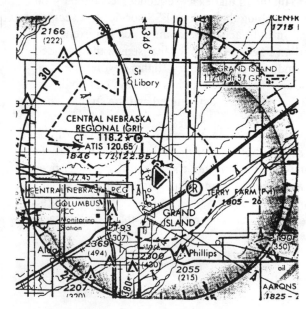

Fig. 5-1. *The ATIS frequency is easy to determine at tower-equipped airports.*

and runway are obtained from the tower. At some of the major airports, two ATIS frequencies are used—one for departing and the other for arriving aircraft, as the Miami excerpt from the *A/FD* (FIG. 5-2) illustrates. These dual transmissions make for shorter messages and reduce the likelihood of pilot confusion.

```
MIAMI INTL    (MIA)    8 NW    UTC-5(-4DT)    25°47'34"N 80°17'26"W                         MIAMI
   11    B    S4    FUEL 100, JET A    OX 1, 2, 3, 4    AOE    ARFF Index E                 H-5E, L-19A, A
RWY 09R-27L: H13002X150 (ASPH-GRVD)    S-130, D-210, DT-420 DDT 850    HIRL CL                 IAP
   RWY 09R: MALSR. VASI(V4L)—GA 3.0°TCH 57'. Thld dsplcd 1352'. Railroad. Rgt tfc.
   RWY 27L: MALSR. VASI(V4L)—GA 3.0° TCH 50'. Thld dsplcd 1270'.
RWY 09L-27R: H10502X200 (ASPH-GRVD)    S-130, D-210, DT-420 DDT 850    HIRL CL
   RWY 09L: ALSF1 TDZ. VASI (V4L)—GA3.0°TCH 54'. Tower.
   RWY 27R: MALSR. VASI(V4L)—GA 3.0°TCH 52'. Tree. Rgt tfc.
RWY 12-30: H9355X150 (ASPH-GRVD)    S-130, D-210, DT-420, DDT-850    HIRL CL
   RWY 12: VASI (V4L)—GA 3.0° TCH 55'. Tower.
   RWY 30: VASI (V6L) Upper—GA 3.0° TCH 94.29'. Lower—GA 3.0° TCH 52.7'. Thld dsplcd 939'. Tree.
AIRPORT REMARKS: Attended continuously. Rwy 09L-27R west 4000' ungrooved. Ldg fee. CLOSED to non-engine
   aircraft. Portions of the inner taxiway Northeast of concourse D to Northeast of concourse B not visible from twr.
   Flight Notification Service (ADCUS) available. NOTE: See SPECIAL NOTICE—Simultaneous Operations on
   Intersecting Runways.
WEATHER DATA SOURCES: LLWAS.
COMMUNICATIONS: ATIS Arr 117.1    Dep 119.15    UNICOM 123.0
   MIAMI FSS (MIA) on Kendall-Tamiami Executive. TF 1-800-WX-BRIEF. NOTAM FILE MIA.
Ⓡ APP CON 124.85 (270°-089°) 120.5 (090°-269°) 125.75
   TOWER 123.9 (090°-269°) 118.3 (270°-089°)    GND CON 127.5 (09R-27L-30) 121.8 (09L-27R-12)
   CLNC DEL 135.35 120.35
Ⓡ DEP CON 125.5 (090°-269°) 119.45 (270°-089°)
   TCA: See VFR Terminal Area chart.
```

Fig. 5-2. *Miami has two ATIS frequencies: one for arrivals, one for departures.*

INFORMATION PROVIDED BY ATIS

Assuming no unusual or potentially hazardous conditions, the typical ATIS includes the following data:

- Location
- Information code (phonetic alphabet)
- Time (UTC, "Coordinated Universal Time," often stated as "Zulu")
- Sky conditions (often omitted if ceiling is higher than 5,000 feet, or is stated as "better than five thousand")
- Visibility (often omitted if visibility is greater than 5 miles, or is stated as "better than five miles")
- Temperature and dewpoint
- Wind direction (magnetic) and velocity
- Altimeter setting
- Instrument approach in use
- Current runway in use
- NOTAMs (if any)
- Information code repeated (phonetic alphabet).

A typical example:

Jackson Information Delta. One six four five Zulu weather. Four thousand five hundred broken, visibility five. Temperature four five, dewpoint three eight. Wind two zero zero at one six, gusting to two two. Altimeter three zero zero niner. ILS Runway One Five in use, land and depart Runway One Five. Advise you have Delta.

Depending on the facility, some ATIS tapes include departure and approach frequencies and the clearance delivery frequency. In the event of adverse weather, ATIS broadcasts include AIRMETs and SIGMETs, such as "SIGMET Charlie Two is now in effect." To determine what Charlie Two is all about, you have to contact Flight Service. ATIS does not provide all of the details.

To illustrate some of the additional information that could be included in an ATIS broadcast, let's take this example:

Jackson Information Foxtrot. One three three zero Zulu special observation. Five hundred scattered, estimated ceiling one thousand overcast, visibility five, light rain. Temperature seven zero, dewpoint six two. Wind three two zero at seven. Altimeter two niner niner zero. VOR Delta approach in use. Left traffic. Land and depart Runway Three Five. SIGMETs Charlie Two Niner, Charlie Three Zero, and Charlie Three One now in effect. Numerous trenches along Runway Three Five. Advise you have Foxtrot.

WHEN TO TUNE TO ATIS

Departure: The engine is started; radios and beacons are on. Before any further action, tune to the ATIS frequency and listen. As it is a continuous transmission, listen twice—especially if the details of the first transmission weren't clear. Set the altimeter, and remember the information code.

Arrival: At 20 or 25 miles out, tune to the ATIS frequency and listen again—perhaps a couple of times so that you have the details firmly in mind. Presumably you're not too busy at that distance from the field, so you can give attention to what's being transmitted and plan your approach to the airport accordingly.

COMMUNICATING THE FACT THAT YOU HAVE THE ATIS

Tell those who need to know that you have the current ATIS that you do indeed have it. This includes Ground Control for departure, and Approach Control or the tower when landing.

To Ground Control (when you're ready to taxi):

You: Jackson Ground, Cherokee One Four Six One Tango at Avitat with Information Foxtrot. Ready to taxi. VFR New Orleans.

GC: *Cherokee One Four Six One Tango, taxi to Runway Three Five.*

You: Roger. Taxi to Three Five. Cherokee Six One Tango.

To the Tower: You're inbound and over an identifiable checkpoint 20 miles or so from the field and this is your first arriving contact with a traffic controlling facility. The initial call, then, is simply this:

Albany Tower, Cherokee One Four Six One Tango over Smithville at three thousand five hundred for landing Albany with India.

Note that your present altitude is included in the call, along with the phonetic designation of the ATIS you have just monitored.

If the airport has an Approach Control, the structure of the message is the same:

1. N-number
2. Location
3. Intentions (landing primary airport, secondary airport, or transiting)
4. ATIS information you have just monitored.

If you fail to indicate that you have the ATIS information, you might well be questioned: "Cherokee Six One Tango, do you have ATIS?" Or the tower or Approach might offer the basic data necessary for your entry into the pattern and for landing. That's a nice gesture, but it consumes airtime.

USING THE PHONETIC DESIGNATION

If you've received the ATIS, make it clear to the controller that you have Information Alpha, Bravo, Charlie, or whatever. That tells the controller that you are in possession of the most current and complete information.

Perhaps it's sort of classy to come out with "have the numbers," but it's misleading. It means to the controller that you have received the wind, altimeter setting, and runway only—not the full ATIS, which might include SIGMETS, NOTAMs, ceilings, visibility, and so on. Hearing "have the numbers" might cause the controller to question you further as to the extent of your information or require him to relay significant information that is included in the latest ATIS.

Another thing: Merely reporting that you "have the ATIS" is not enough. Maybe a special update was just issued. Maybe you had Information Papa 20 minutes ago, but Information Quebec has come out in the meantime. And maybe Quebec designates Runway 18 as the active runway, whereas Papa had it as Runway 26. Any number of changes might have taken place between Papa and Quebec.

To eliminate confusion and reduce airtime talk, be brief but be complete. Use the phonetic alphabet to identify the ATIS you have received. It takes no longer to be specific—and it's proper procedure. As the AIM says, " 'Have the numbers' . . . does not indicate receipt of the ATIS broadcast and should never be used for this purpose."

WHEN UNABLE TO REACH THE ATIS

If, for any reason, you're unable to receive the ATIS or the transmission is garbled, tell Ground Control before taxiing out, or tower or Approach Control if you're coming in. Just conclude the call with "negative ATIS" or "ATIS garbled." The facility you're contacting will then give you the pertinent data so that you can proceed accordingly.

CONCLUSION

ATIS broadcasts exist to: 1) provide you with the departure or arrival information you need (at worst, the information is only 59 minutes old, and is thus a reasonably accurate depiction of existing conditions), and 2) reduce unnecessary communications, thereby permitting pilots and controllers to concentrate on their primary responsibilities—the safe operation of the aircraft and the safe direction of air traffic.

Use ATIS wherever it is available. You don't have to talk back to it. All it requires is the willingness to listen and the ability to remember. If you can't do either of those, you shouldn't be in the air in the first place.

6
Ground Control

Most pilots who inhabit the busier airports are well acquainted with Ground Control and the function it plays in assuring the safe operation of earthbound traffic. However, the inexperienced airman, venturing into one of these airports for the first time, needs to understand the responsibilities of Ground Control and the communications with the controller. And so do some of the more experienced aviators. Let's just say that I've heard a lot of hesitant radio calls, incomplete calls, unprepared calls, and excessive verbal garbage—all of which reflects poorly on those who should know better.

What follows, then, should be of some value to all general aviation pilots, regardless of their levels of experience.

WHAT DOES GROUND CONTROL DO?

Ground Control is the invisible "policeman" ensconced in the tower who is responsible for the operation of aircraft and all other vehicles that are using taxiways and runways other than the active runway. It controls the movement of those vehicles through radio communications. The tower is the authority. When the controller says "go," you go; when he says "stop," you stop; when he says "give way," you give way, and so on. If, for any reason, safety or otherwise, you can't comply with his directions, you have the responsibility to advise the controller accordingly. Adherence to his instructions is expected.

At almost every airport, however, some areas are not under the control of Ground: for example, the expanses of concrete used for tiedown purposes, fueling, moving aircraft to and from hangars, compass swinging, and the like. As long as these areas don't infringe on runways or taxiways, you can taxi your aircraft around there all day. Nobody will say a word. Otherwise, you must have your intended movements cleared by Ground Control.

FINDING THE GROUND CONTROL FREQUENCY

The Ground Control frequency will not be found on any sectional. Where the service exists (at most tower-controlled fields), you can locate the frequency in the *Airport/Facility Directory*, as FIG. 6-1 illustrates.

ATLANTA

DEKALB-PEACHTREE (PDK)   8 NE   UTC–5(–4DT)   33°52'30"N 84°18'08"W   **ATLANTA**

1002   B   S4   **FUEL** 100, JET A   OX 1, 3   TPA (See Remarks)   LRA   **H-4H, 6F, L-20E, A**

RWY 02R-20L: H6001X100 (CONC–GRVD)   S-46, D-66, DT-90   HIRL   0.3% up S.   **IAP**

RWY 02R: REIL. VASI(V4L)—GA 3.9°TCH 35'. Building. Rgt tfc.   RWY 20L: MALSF. Thld dsplcd 1000'. Trees.

RWY 16-34: H3966X150 (ASPH)   S-20   MIRL

RWY 16: RAIL. REIL. ODALS. VASI(V4L)—GA 3.4°TCH 30'. Building.

RWY 34: REIL. VASI(V4L)—GA 3.3°TCH 39'. Trees.

RWY 02L-20R: H3744X150 (ASPH)   S-20   MIRL

RWY 02L: VASI(V2L)—GA 3.5°TCH 39'. Trees.   RWY 20R: VASI(V2L)—GA 4.0°TCH 42'. Rgt tfc.

RWY 09-27: H3378X150 (ASPH)   S-20   HIRL

RWY 09: REIL. VASI(V4R)—GA 3.4°TCH 28'. Trees.   RWY 27: REIL. VASI(V4L)—GA 3.8°TCH 49'. Trees.

AIRPORT REMARKS: Attended continuously. Heavy helicopter opr N.W. corner of arpt. Flocks of birds on or near arpt during dalgt hours. ACTIVATE ODALS Rwy 16 and REIL Rwy 16–123.0. When twr clsd HIRL Rwy 09–27 and MIRL Rwy 16–34 available by PPR ctc—120.9 or phone 404-451-1193. When twr clsd ACTIVATE MALSF Rwy 20L—120.0. ACTIVATE ODALS Rwy 16—123.0. TPA—2002 (1000) single engine, 2502 (1500) multi engine. Voluntary ngt curfew in effect from 0400-2200Z‡. Noise sensitive area all quadrants; pilots use noise abatement procedures prescribed by arpt director, call 404-457-7236. Flight Notification Service (ADCUS) available. Control Zone effective1130-0400Z‡ Mon-Fri, 1200-0400Z‡ Sat-Sun. NOTE: See SPECIAL NOTICE— Simultaneous Operations on Intersecting Runways.

WEATHER DATA SOURCES: LAWRS.

COMMUNICATIONS CTAF 120.9   ATIS 128.4   UNICOM 122.95

ATLANTA FSS (ATL) LC 691-2240. NOTAM FILE PDK.

ATLANTA APP/DEP CON 119.3

PEACHTREE TOWER 120.9 120.0 (1130-0400Z‡ Mon–Fri 1200–0400Z‡ Sat–Sun)   **GND CON 121.6**

CLNC DEL 125.2

RADIO AIDS TO NAVIGATION: NOTAM FILE ATL.

PEACHTREE (L) VOR/DME 116.6 PDK Chan 113   33°52'32"N 84°17'56" at fld. 970/02W.

ILS 111.1 I-PDK Rwy 20L. Unmonitored when twr clsd.

ASR

Fig. 6-1. *How the* A/FD *identifies the Ground Control frequency.*

One fact you can usually count on: the frequency will be 121.6, 121.7, 121.8, or 121.9. There are a few exceptions to the general rule, however, as at Detroit Metro. (FIG. 6-2). Miami International uses one frequency for one set of runways and another frequency for the other runways. Memphis has one frequency for general aviation, while a second is reserved for air carriers. To be sure of your Ground frequency, consult the latest edition of the *A/FD*.

§ **DETROIT METROPOLITAN WAYNE CO** (DTW) 15 S UTC–5(–4DT) 42°12′55″N 83°20′55″W **DETROIT**
639 B S4 **FUEL** 80, 100LL, JET A, A1 + OX 1, 3 AOE CFR Index E **H-3C, L-23C, A**
RWY 03L-21R: H10501X200 (CONC-GRVD) S-100, D-185, DT-350 HIRL CL **IAP**
  RWY 03L: ALSF2. TDZ. Tree.   **RWY 21R:** MALSR. Railroad.
RWY 03R-21L: H10000X150 (CONC-GRVD) S-100, D-200, DT-350, DDT-750 HIRL CL
  RWY 03R: ALSF2. TDZ.   **RWY 21L:** MALSR.
RWY 09-27: H8700X200 (ASPH-CONC-GRVD) S-100, D-185, DT-350 HIRL
  RWY 09: REIL. VASI(V4R)—GA 3.0° TCH 46′.   **RWY 27:** MALSR. Tree.
RWY 03C-21C: H8500X200 (ASPH-CONC-GRVD) S-100, D-185, DT-350 HIRL
  RWY 03C: REIL. VASI(V4L)—GA 3.0° TCH 60′. Tree.   **RWY 21C:** REIL. VASI(V4L)—GA 3.0° TCH 59.4′.
AIRPORT REMARKS: Attended continuously. Landing fee. Rwy 03R ALSF2 required when RVR/visibility is 6000/1 mile or less. SSALR ops when RVR/visibility is 6000/1 mile. Flight Notification Service (ADCUS) available.
WEATHER DATA SOURCES: LLWAS.
COMMUNICATIONS: ATIS 124.55 (313) 942-9350   **UNICOM** 122.95
  LANSING FSS (LAN) Toll free call, dial 1-800-322-5552. NOTAM FILE DTW
  CARLETON RCO 122.1R, 115.7T (LANSING FSS)
Ⓡ **DETROIT APP CON** 124.05, 124.25 (211°-029°) 125.15 (030°-210°)
  METRO TOWER 135.0 (WEST) 118.4 (EAST) **GND CON** 121.8 (WEST) 119.45 (EAST)   **CLNC DEL** 120.65
  PRE TAXI CLNC 120.65
Ⓡ **DETROIT DEP CON** 120.15 (030°-210°) 118.95 (211°-029°)
TCA Group II: See VFR Terminal Area Chart.
RADIO AIDS TO NAVIGATION: NOTAM FILE LAN
  CARLETON (H) VORTAC 115.7 CRL Chan 104 42°02′53″N 83°27′28″W 033° 10.7 NM to fld. 630/3W.
  WILLOW RUN (T) VORW/DME 110.0 YIP Chan 37 42°14′13″N 83°31′31″W 102° 7.3 NM to fld. 707/3W.
  REVUP NDB (LOM) 388 DT 42°07′12″N 83°25′55″W 037° 6.4 NM to fld.
  SPENC NDB (LOM) 223 DM 42°13′12″N 83°12′13″W 273° 5.7 NM to fld.
  ILS/DME 110.7 I-DTW Chan 44 Rwy 03L LOM REVUP NDB
  ILS/DME 110.7 I-DWC Chan 44 Rwy 21R
  ILS 111.5 I-HUU Rwy 03R
  ILS 111.5 I-EJR Rwy 21L.
  ILS 108.5 I-DMI Rwy 27 LOM SPENC NDB
  ASR
COMM/NAVAID REMARKS: Willow Run VOR/DME out of svc indefinitely.

Fig. 6-2. *Detroit Metro is an example of dual Ground Controls: one frequency for the west side of the field, another for the east.*

WHEN TO CONTACT GROUND CONTROL

The following are typical situations in which you would contact Ground Control. Some contacts are required, and others are optional, as noted.

Taxiing for Takeoff (Required)

You have the ATIS information, you're tuned to the correct Ground Control frequency, and are ready to taxi out now—not one, three, or five minutes from now. You can roam around the ramp or any uncontrolled area all you want, but don't venture out onto any taxiway until you have contacted Ground and have received permission to taxi.

You: Lincoln Ground, Cherokee One Four Six One Tango at the terminal with Information Lima. VFR Omaha.

GC: *Cherokee One Four Six One Tango, taxi to Runway Three Five.*

You: Roger. Cherokee One Four Six One Tango.

In this contact, be sure to include your location (e.g., " . . . at the terminal . . ."). If you don't, Ground will always come back with something akin to, "Cherokee Six One Tango, where are you parked?" Avoid this needless request by communicating your position in the first call. By including your destination or direction of travel (which really isn't essential), you might be directed (winds permitting) to a runway more closely aligned with your departure route.

Note that in the dialogue the controller has told you to taxi *to* Runway Three Five. That means exactly what it says. You can go to the engine runup area, stop there for the cockpit check and runup, and then taxi to the runway hold line—but no farther. Clearance to taxi onto the active runway for takeoff comes from the tower, not Ground. No further clearance is necessary, however, to taxi from the runup area to that hold line. The fact that Ground has approved your movement to Three Five is all that is required. Another thing: While you're moving down the taxiway, once cleared, keep listening to Ground. The controller might have further instructions for you, such as:

> **GC:** *Cherokee One Four Six One Tango, pull to the right and give way to the Bonanza taxiing in on Charlie (the taxiway).*
>
> **You:** Roger, will do, Six One Tango.

Or:

> **GC:** *Cherokee One Four Six One Tango, turn left at the next intersection to let the 727 pass.*
>
> **You:** Roger, Ground, will do. Six One Tango.

Something else to keep in mind at large or busy airports is that the Ground Controller might lose track of you. In a peak traffic period, for example, he tells you to ". . . follow Charlie to Runway Two Seven and hold." This you do—and you hold and you hold and you hold, awaiting further word from Ground. But the controller has either lost you or has forgotten about you amidst the demands from other ground traffic. The practical thing, then, is to tell Ground when you've reached "Two Seven":

> **You:** Ground, Cherokee Six One Tango holding at Two Seven.

This does two things: First, it reattracts the controller's attention to you, and second, it advises him that you've followed his instructions. Controllers like that, especially the latter. Then when he clears you across Two Seven, be sure to acknowledge the approval:

> **GC:** *Cherokee One Four Six One Tango, clear to cross Runway Two Seven.*
>
> **You:** Roger, clear to cross Two Seven.

Soliciting Progressive Taxi Instructions (Optional)

You have the ATIS, but you're new to the airport and need taxi instructions. Ground controllers understand that taxiway systems can be confusing to pilots, so they will give progressive (step-by-step) taxi instructions, if you ask for them:

You: Lincoln Ground, Cherokee One Four Six One Tango at the terminal with Information Lima. VFR Omaha. Request progressive taxi.

GC: *Cherokee Six One Tango, Roger. Taxi to the edge of the ramp and hold.*

You: Roger, edge of the ramp and hold. Cherokee Six One Tango.

After proceeding as told, come to a stop, acknowledge your position, and wait for Ground to come back with further instructions.

You: Ground, Cherokee Six One Tango holding at ramp edge.

GC: *Cherokee Six One Tango, you are at Taxiway Charlie. Turn right on Charlie to Runway Two Seven and hold.*

You: Ground, Cherokee Six One Tango holding Two Seven.

GC: *Cherokee Six One Tango, cleared to cross Two Seven. Turn left at the next intersection to Delta. Follow Delta to Runway One Five and hold. Contact the tower when ready.*

You: Six One Tango cleared to cross Two Seven, left on Delta to One Five and hold. Cherokee Six One Tango.

Just don't be afraid to call Ground again if you've forgotten an instruction or become confused. The folks in that glass enclosure are there to help you and to keep things running smoothly on those strips of concrete.

Taxiing in after Landing (Required)

You've touched down, and the tower has instructed you to "contact Ground point niner." (The tower might give you the complete frequency, as "121.9," but more likely will shorten the instruction to "point niner.") When clear of the active runway—and past the hold line—come to a full stop and change to the Ground frequency:

You: Lincoln Ground, Cherokee One Four Six One Tango is clear of Runway Three-Five (or "the active"). Request taxi to the terminal. [Or you can request progressive taxi instructions as above.]

GC: *Cherokee Six One Tango, taxi to the terminal.*

You: Roger. Cherokee Six One Tango.

Remember to stay tuned to the Ground frequency any time you're moving the aircraft in a controlled area. Yes, you've landed and you've been cleared to taxi to the terminal, but just as in the taxiing-out example, Ground might need to give you subsequent instructions. This admonition is probably superfluous, but pilots have been known to switch everything off—rotating beacon, radios, transponder, and so forth—once cleared to the parking area. Except for the transponder, that's a no-no to those responsible for the control of airport ground traffic.

Moving the Aircraft from One Ground Location to Another (Required, Except on Ramp/Uncontrolled Areas)

Let's say you need to taxi to the other side of the field for a minor repair at a radio shop. To do so, you have to use a taxiway and cross the active runway. Permission for both is mandatory:

You: Lincoln Ground, Cherokee One Four Six One Tango at the terminal. Request taxi to King Radio.

GC: *Cherokee Six One Tango, taxi on Foxtrot. Hold short of Runway One Eight.*

You: Roger, hold short of Runway One Eight. Cherokee Six One Tango.

GC: *Cherokee Six One Tango, cleared to cross Runway One Eight.*

You: Roger. Cleared across One Eight. Cherokee Six One Tango.

Getting a Current Altimeter Setting (Optional)

You're ready to taxi from the ramp. After listening to the ATIS, however, you find that the altimeter setting given in the recording doesn't coincide with the field elevation. Realizing that the ATIS might be almost an hour old, you want the current reading. Incorporate your request for this data in the initial call to Ground:

You: Lincoln Ground, Cherokee One Four Six One Tango at the terminal with Information Lima. Ready to taxi. VFR Omaha. Request altimeter setting.

GC: *Cherokee One Four Six One Tango, Lincoln altimeter is three zero zero five. Taxi to Runway Three Five.*

You: Roger, three zero zero five. Taxi to Three Five. Cherokee Six One Tango.

For a Radio Check (Optional)

You're on the ramp and want to determine the clarity and volume of your transmission. The easy way to do this is to call Ground and ask for a radio check:

You: Lincoln Ground, Cherokee One Four Six One Tango radio check.

GC: *Cherokee Six One Tango, loud and clear.*

You: Roger, Cherokee Six One Tango.

If Ground comes back with "transmission is scratchy and volume weak," don't indulge in an explanatory harangue. You've got a sick radio, Ground has told you so, so get off the air, and have a technician look at it. Ground can't help you a bit!

IS THE AIR CLEAR?

Now is as good a time as any to stress the point of listening before you speak. In other words, is the air clear?

You're at the ramp and have just tuned to Ground for taxi permission. The first thing that you hear is an IFR aircraft requesting its clearance. Ground responds with:

GC: *Cessna Three Four Romeo is cleared as filed to Denver. Maintain three thousand, expect ten thousand fifteen minutes after departure. Departure Control one two six point six. Squawk zero three two five. Fly heading two four zero after departure.*

[A brief period of silence]

34R: Cessna Three Four Romeo cleared as filed. Three thousand, ten thousand in fifteen. Departure one two six point six, zero three two five, two four zero heading.

GC: *Cessna Three Four Romeo, readback correct. Taxi to Runway One Eight.*

34R: Roger, taxi to One Eight. Cessna Three Four Romeo.

As indicated in brackets, there will almost always be a pause, a brief period of silence after Ground has conveyed the clearance, while the pilot is copying the instructions. Then comes the readback.

Don't start transmitting just because the air is momentarily quiet. Recognize what's taking place and give Three Four Romeo time to complete its task and reestablish communications with the controller. This courtesy applies to all communications situations, short of an emergency. Give both parties the opportunity to acknowledge instructions, answer a question, repeat an instruction, and the like. Momentary silence doesn't necessarily mean the air is all yours.

Pilots who don't listen and aren't considerate are usually the reason for the squeals and squeaks that distort reception. Two people can't transmit at the same time on the same frequency without creating that discordant cacophony that penetrates the cockpit or headset.

CONCLUSION

Ground Control is the "policeman" for all ground traffic—cars, trucks, tugs, and aircraft. Its use is mandatory. Even more than that, however, it is a source of assistance and an overseer of safety, ensuring the smooth flow of ground operations. Additionally, its very existence enables the tower, responsible for the smooth flow of flight operations, to concentrate solely on that responsibility.

It thus behooves all pilots to be familiar with what Ground Control can do for us. You must use the service, but you should use it wisely by being clear and concise and observing the basic rules of courtesy. The Ground Controller will invariably respond in kind.

7
Transponders

While I have referred to transponders in some of the previous discussions and examples, this piece of electronic hardware hasn't been a major factor in the communication processes up to now. Henceforth, however, as I go through the sections on tower, Approach and Departure Control, and the Air Route Traffic Control Centers, the importance of the transponder increases—as does the importance of the terminology associated with it. Consequently, if you own or rent a transponder-equipped aircraft, some familiarity with it is in order.

THE AIR TRAFFIC CONTROL RADAR BEACON SYSTEM (ATCRBS)

Simply said, the basic radar system is composed of two elements. One is the primary radar, which scans the surrounding area and identifies on the radarscope, or screen, objects such as buildings, radio towers, aircraft without transponders, and aircraft with transponders turned off.

As this relatively minimal identification has safety and traffic control limitations, a secondary radar system was developed which incorporates a ground-based transmitter-receiver called an *interrogator*. This system—the Air Traffic Control Radar Beacon System (ATCRBS)—functions in unison with the primary radar and, in the scanning process, "interrogates" each operating transponder. In effect, it "asks" the transponder to reply. The primary and secondary signals are then synchronized and together transmit a distinctly shaped blip or target to the controller's radar screen.

That target, however, only tells the controller that there's an aircraft out there with a transponder "squawking" the standard VFR code of 1200. It does not permit more specific identification of the aircraft, which could be important in periods of heavy traffic or poor visibility.

Consequently, each transponder is equipped with an identification feature: the IDENT button. When the button is pushed, the radar target changes shape to distinguish the identing aircraft from other aircraft on the controller's screen. Very broadly and nontechnically, this is the radar beacon system. For those interested in more detail, make a visit to a radar-equipped tower or an Air Route Traffic Control Center. The specialists in either location are always glad to explain the system and let you watch it in operation—workload permitting, of course.

THE TRANSPONDER: Types (or modes)

You'll often hear references to transponder "types" or "modes." In the event there's any uncertainty as to what those references mean, let's take a moment to clear the air.

Several different controller-operator decoder modes exist. *Mode 1* and *Mode 2* are reserved for the military. *Mode 3/A* is common to both civil and military use. *Mode B* applies to traffic of foreign countries. *Mode C* identifies a transponder that is equipped with altitude-reporting capabilities. *Mode D* is not currently in use. *Mode S* will be the standard in the mid- to late-1990s when collision-avoidance systems are due to be implemented. (Mode S will automatically transmit the aircraft's N-number, type, and altitude—eliminating the need to dial in and squawk different codes. It will also permit onboard computer terminals to communicate with ground facilities, resulting in inflight printouts of clearances, weather charts and forecasts, and so on. Currently in United States civilian flying you use Mode 3/A and Mode C.

The transponder illustrated at the beginning of this chapter (a Narco AT 50A) has five switch positions: OFF, SBY (standby), ON, ALT (altitude), and TST (test). Additionally, there are four code selector knobs so that the pilot can dial in whatever four-digit numerical code Air Traffic Control requests. Each of the four knobs can bring up numbers from 0 to 7, thus allowing for a total of 4,096 ($8 \times 8 \times 8 \times 8$) separate or discrete codes, hence the frequent reference to a "4096 transponder."

Finally, the transponder is equipped with a small reply light that blinks every time the transponder responds to the radar beacon interrogator. These blinks also confirm to the pilot that the transponder is functioning. Associated with the light is an ident feature. On the unit illustrated, the ident button and the reply light are one and the same. On other units, the button might be separate. If ATC asks you to "Ident," you merely push the button, and the image on the controller's radar screen changes. If you're one of only a few aircraft in the same general area, the controller can track you fairly easily once you and the others have "idented." It's more difficult for him, though, if the activity is heavy. In these cases, he might ask you to report when over a certain landmark for verification of your position.

What I've said describes the basic Mode 3/A unit. What Mode 3/A doesn't have is the altitude-reporting capability of Mode C. The 3/A is converted to Mode C merely by adding to the system either an *encoding altimeter* or a *blind encoder*. The first is a normal-appearing altimeter that is coupled to the transponder. As the altimeter aneroid bellows expand and contract with pressure changes, these changes are converted to coded response pulses by the transponder, which then transmits the aircraft's altitude (to the nearest 100 feet). The blind encoder performs the same function, but the unit is usually located on the firewall, out of the pilot's sight.

Of the two, the blind encoder has certain advantages over the panel-mounted encoding altimeter. For one, it's usually a little less expensive and can be installed at about the same cost. Also, if it fails, it can be removed for repair and the aircraft operated as usual, except where Mode C is required.

On the other hand, if the encoding altimeter goes out and you still want or need altitude-reporting capability, the entire altimeter has to be removed and the aircraft is grounded pending repairs. Finally, installing an encoding altimeter means the replacement of what is probably a perfectly functioning unit. If you can sell the unit, fine. Otherwise, you've got a good altimeter to put on your fireplace mantle or office desk to impress visitors.

TRANSPONDERS: Where and when required

With the concern for inflight safety heightened by recent midair collisions and reported near-misses, more stringent operating and equipment requirements have been established: Barring certain exceptions listed at the end of this section, it's almost mandatory today that every aircraft be equipped with an operable transponder. Otherwise, as the following indicates, freedom of flight is severely limited.

- The transponder must be in the ON position, or ALT position if Mode C, whenever you are operating in controlled airspace (which is just about everywhere in the U.S.).

- Mode C is required at and above 10,000 feet msl to the floor of the Positive Control Area (18,000 feet msl), excluding the airspace at and below 2,500 feet agl.

- An operating Mode C transponder is further required within 30 nautical miles of a Terminal Control Area's (TCA) primary airport, from the surface up to 10,000 feet msl. (Note: clarification of this regulation follows in a moment.)

- A Mode C transponder is also required within an Airport Radar Service Area (ARSA) airport, up to and including 10,000 feet msl.

- Finally, Mode C is required within a 10-nautical-mile radius of a few airports that have been classified as "Designated." These are airports which are neither TCAs nor ARSAs but which, for particular operational reasons, justify the control that Mode C permits. To date, the only two airports so affected are

Logan International Airport, Billings, Texas, and Hector International Airport, Fargo, North Dakota.

Exceptions to these regulations are aircraft not originally certificated with an engine-driven electrical system, balloons, and gliders, as long as they operate outside any terminal control area, positive control area, and below 10,000 feet msl.

Now relative to the clarification noted above:

If you check any sectional with a Terminal Control Area airport, you'll see a thin blue circle at the outer extremes of the TCA. This circle represents the 30-nautical-mile radius, or "Mode C veil," surrounding the primary TCA airport.

According to the initial regulation, all aircraft operating into or out of any airport within that circle had to have Mode C transponders. That raised a storm of protest, however, especially from many smaller, and, more often than not, uncontrolled airports that were right on the 30-nm circle or only a mile or so inside of it.

As a result of the protests, the FAA modified the regulation by excluding from the Mode C regulation those airports that are within two nautical miles of the Mode C veil and from which direct flights can be made to the outer boundary of the veil (in other words, to and beyond the blue circle). One further stipulation is that operations must be conducted below the maximum altitude established for those airports—altitudes that vary with the individual TCAs but are generally 1,000 up to 2,500 feet agl.

Exclusion from the regulation does not mean that non-Mode C aircraft can operate inward toward the TCA airport. They can take off, land, or practice touch-and-gos in the traffic pattern at those airports as long as they remain below the maximum altitude. Otherwise, they are expected to leave the veil as expeditiously as possible after takeoff.

All told, there are approximately 300 Mode C-excluded airports, so if you're planning a trip to one that is right on the fringe of the veil (per the sectional), I suggest that you check the *Airman's Information Manual*. *AIM* has a complete listing of the exclusions across the country immediately under the TCA in question, along with the airport ID and the maximum altitude at which operations can be conducted.

All of these regulations do limit the airspace for non-Mode C aircraft, especially near urban areas, but Mode C has a lot of advantages, not the least of which is the added safety it provides. Plus that, it should keep pilots more alert to what they're doing.

For example, if you accidentally or intentionally penetrate a TCA without approval, you can be sure that someone on the ground knows it. Your position, combined with your altitude readout from the Mode C, gives you away. Oh, you might be able to outrun the radar coverage and get home unidentified, but the FAA is continually developing more sophisticated means of tracking airspace violators. And once caught, it could mean a license suspension of 60 days or longer, plus a black mark on your record. With Mode C in operation, though, you know that you're probably being monitored. That ought to have a powerful effect on regulation abidance and attention to what you're doing.

TRANSPONDER OPERATION AND CODES

Operating the transponder is simply a matter of positioning the activating switch, being aware of the use of the Ident button, and, as I'll discuss shortly, entering the codes. A couple of hints about the operation, however, might be in order:

STANDBY: After engine start-up, turn on the radio(s) and put the transponder switch to the SBY position to allow the set to warm up without replying to the radar interrogator. Keep the switch in this position until cleared for takeoff or when you have actually begun the takeoff roll.

ON or ALT: With Mode 3/A, switch from SBY to the ON (or in some units, NORMAL) position when cleared for takeoff, and leave it there throughout the flight, unless directed otherwise by ATC. If the unit has Mode C, switch to the ALT position. This not only turns the set on, but now it also reports your altitude. As mentioned earlier, if you have a transponder when operating in controlled airspace (unless ATC directs otherwise), the transponder must be switched to ON (Mode 3/A), or, if equipped with an altitude encoder, ALT (Mode C).

OFF: Turn the transponder off as soon as you have landed, either on the rollout or when you're clear of the runway. If you leave it on, all you're doing is painting an enhanced image on the screen because of your proximity to the radar beacon antenna.

IDENT: Only when ATC asks you to "Ident" do you push the IDENT button—and just once. The signal sent will change the shape of the blip on the screen. Now the controller can more readily identify your aircraft and its relation to ground obstacles and other airborne traffic.

Above and beyond the basic operation of the transponder, it's essential that pilots be familiar with the standard numerical codes that are controlled by the four knobs on the set. The most common code is 1-2-0-0, which is the standard for all VFR altitudes and operations. There are exceptions to that statement, though, because if you are in contact with a Center or Approach or Departure Control, the controller will tell you to enter some other code (called a *discrete code*) for his specific identification of your aircraft.

Let's amplify that a bit. You've started out on a cross-country with the 1200 VFR code in the transponder. Then you call a Center and ask for "enroute traffic advisories." As a lot of VFR aircraft could be out there, the controller wants to spot your particular aircraft, so—and only for example—he asks you to "squawk 2056." This, or whatever four-digit combination he gives you, is a discrete computer-generated code that is assigned only to you. Once you have entered the new code, the computer recognizes it and displays your aircraft with a VFR symbol, your N-number, ground speed and, if Mode C-equipped, your altitude on the controller's screen. Now he has you distinguished from all other aircraft and is better able to alert you to possible conflicting traffic.

Other Codes

Along with 1-2-0-0, a few other standard codes have been established, as summarized in TABLE 7-1. Note, though, that the codes with asterisks are never to be used by civilian pilots.

Table 7-1. Transponder codes

Code	Type of flight	When used
0000*	Military only	North American Air Defense
1200	VFR	All altitudes unless otherwise instructed
4000*	Military VFR/IFR	Warning/Restricted areas
7500	VFR/IFR	Hijack
7700	VFR/IFR	Emergency ("Mayday")
7700 (1 min.) 7600 (15 mins.)	VFR/IFR	Loss of radio communications
7777*	Military only	Interceptor operations
Assigned by ATC	VFR/IFR	When using Center or Approach/Departure Control

TERMINOLOGY

Next, the terminology, or phraseology, associated with Mode 3/A and Mode C transponder operation:

Squawk: The origin of this rather odd term goes back to World War II and a radar beacon system called IFF (Identification, Friend or Foe). Allied aircraft were equipped with transmitters that replied to radar sweeps with a sound similar to a parrot's squawk. Today the term is used by controllers and pilots alike to indicate that the transponder is on and that a certain four-digit code should be, or has been, dialed in. If a controller asks you to "squawk two zero five six" (or any code), he wants you to enter those digits in the transponder. That is not an instruction to "ident" however. If you're squawking a code other than 1200 and are told to "squawk VFR" it means to change the code back to 1200.

Ident: When a controller says, "Cherokee Six One Tango, ident," he wants you to push the little IDENT button so that he can more positively identify your aircraft. But push the button only once and only momentarily. Usually, but not always, after the radar target changes shape, the controller will come back to you with, "Cherokee Six One Tango, radar contact," or "Ident received." Then it's appropriate to acknowledge with, "Roger, Cherokee Six One Tango." On the other hand, when asked to ident, you don't have to reply with, "Roger, Cherokee Six One Tango identing." Just push the ident feature and say nothing. The blip change on the screen is acknowledgment enough for most controllers.

Occasionally, ATC may come back with, "Cherokee Six One Tango, I did not receive your ident. Ident again." In this case, acknowledgment is in order: "Roger, Cherokee Six One Tango." Now repeat the ident.

Squawk (number) and Ident: This instruction asks you to insert a certain numerical code and push the IDENT button. The word "and" might or might not be included in the instruction.

Stop Squawk: When you hear this, it means that the controller wants you to turn the transponder to OFF.

Squawk Standby: This instruction tells you to turn the switch from ON or ALT to SBY, the standby position. Remember, the transponder is not off. It's still warm but isn't responding to any interrogation and thus not transmitting. Your aircraft will be reflected on the screen only by the primary radar—not the secondary. (You might recall the pilot in Cherokee 41966 back in chapter 1. He switched the transponder to BY, not understanding that the controller wanted him to squawk 0252 and then standby for further instructions.)

Stop Altitude Squawk: If you have Mode C and hear this, merely switch from the ALT to the ON position. You're now functioning in Mode 3/A only, which provides aircraft identification but not altitude.

Squawk Mayday: You've verbally communicated an emergency to ATC, and for positive identification, the controller wants you to change your code to 7700.

THE TRANSPONDER AND RADIO COMMUNICATIONS IN AN EMERGENCY

You're cruising along when suddenly—or gradually—things begin to go wrong. Maybe it's a coughing engine, a dead engine, a fire, pilot illness, low fuel, you're lost, or whatever. In any event, action is necessary now, either to combat the immediate emergency or to prevent a potentially serious situation from becoming really serious. So, in the context of this subject, what should you do?

First, "Emergency" Definitions

Two terms are used to describe the nature of an emergency: One is *Distress*, while the other is *Urgency*. A *distress* condition is one involving a fire, a mechanical failure, or a structural failure. In other words, it's a condition that demands immediate attention and immediate assistance. An *urgency* is not necessarily perilous at the moment but could become potentially distressful. Examples: a low fuel supply, you're lost, heavy icing, pilot illness, weather, an unnatural engine vibration, or the like. In effect, an urgency warrants ground assistance, but the situation has not yet reached distress proportions.

Transponder Operation

When a distress or urgency situation arises, pilots with a coded aircraft transponder should immediately enter the 7700 emergency squawk and then attempt to establish radio communications with an ATC facility.

The 7700 code appears on the screens of all radar-equipped facilities within range of their radar coverage, and, by sound as well as a flashing blip on the screen, attracts the controller's attention. Even if not in radio contact with the aircraft, the facility, or facilities, is alerted to the existence of an aircraft in trouble. ·

Radio Communications

If you've been in contact with a tower, a Center, or an FSS when either a distress or urgency develops, the communications should obviously be directed to that facility. A distress message should start with the word "Mayday" and be repeated at least three times. ("Mayday" comes from the French, "M'aidez," meaning "Help me.") In an urgency, the word "Pan-pan," again repeated three times, starts the radio transmission.

Distress messages have absolute priority over all others, and the word "Mayday" commands radio silence on the frequency in use. Urgency messages have priority over all others except distress. "Pan-pan" thus warns other stations not to interfere with urgency transmissions. That noninterference applies to pilots as well. If you hear either "Mayday" or "Pan-pan," stay off the air so that the ground facility can have uninterrupted communications with the troubled aircraft.

If you have not been receiving services from a facility, such as a Center or a tower, and, if time permits, the Center, tower, or FSS in whose area of responsibility the aircraft is operating should be contacted on the appropriate frequency. By so doing, a more rapid response to the call is likely—a good reason for listing the various enroute frequencies in the preflight planning and having the list immediately available in the cockpit.

Conditions might be such, though, that you have no time to search for the correct frequency, even with a prepared list. In such cases, two other frequencies can be used: the *emergency-only* frequencies of 121.5 or 243.0. Both have ranges generally limited to line of sight. The frequency 121.5 is guarded by direction-finding stations and some military and civil aircraft. Frequency 243.0 is guarded by military aircraft. In addition, both frequencies are guarded by military towers, most civil towers, FSSs, and radar facilities.

The most logical facility to try to reach first on 121.5 is the nearest ARTCC. The ARTCC's emergency frequency capability, however, does not normally extend to its radar coverage limits, so if you get no response, address the call to the nearest tower or FSS or to "any station . . . ."

Recognizing the problem of time in a distress condition, as much of the following as possible, preferably in the order suggested, should be communicated:

1. "Mayday, Mayday, Mayday" or "Pan-Pan, Pan-Pan, Pan-Pan"
2. Name of facility addressed or "any station"
3. Aircraft identification and type
4. Nature of distress or urgency
5. Weather
6. Pilot's intentions and request
7. Present position and heading, or if lost, last known position, time, and heading since that position
8. Altitude or flight level
9. Hours and minutes of fuel remaining
10. Any other useful information, such as visible landmarks, aircraft color, emergency equipment on board, number of people on board
11. Activate the Emergency Locator Transmitter (ELT), if installation permits.

As an illustration of a distress call: Assume that you're about 50 miles east of Atlanta on Victor Airway 18, destination Columbia, South Carolina. Along the way, you've been monitoring Atlanta Center to get an idea of the traffic in the area, but you haven't been in contact with the facility for traffic advisories. Suddenly you have an oil line break. The pressure gauge drops rapidly and the spurting oil covers the windshield with a coat of film that reduces forward visibility almost to zero. You've got a problem.

The first thing to do is enter 7700 in the transponder. Then get on the air to Atlanta Center—Atlanta because the frequency is already tuned in and no time is wasted changing to 121.5. Even though you haven't been talking to the Center, the fact that you're listening on the frequency means that you can establish immediate contact with the controller responsible for your sector or area.

Now an example of the call—a call that is as concise as possible, communicating only the most essential information.

You: Mayday, Mayday, Mayday. Atlanta, Cherokee One Four Six One Tango distress. Oil line break. No forward visibility account break. Land immediately. Fifty east, Victor 18 at six point three (that's pilot jargon for "six thousand three hundred" feet altitude), descending. White, red stripes. Two aboard.

Note that items 5 and 9 of the information listed above have not been addressed in this example. They're omitted because the weather is presumably VFR or you wouldn't be there in the first place, and fuel remaining is hardly consequential in a mechanical emergency. If you were lost or low on fuel, both elements would be important to people on the ground, but not here.

Note, too, that there's no unnecessary verbiage. Speak as calmly and clearly as possible, but make the messages short, terse, to the point.

I won't simulate further exchanges, because the main burden for instructions now falls on the controller. Your job is: 1) to maintain control of the airplane; 2) to comply, to the best of your ability, with his instructions; and 3) to keep the controller informed of what's taking place.

Pilot Responsibilities after Establishing Radio Contact

Once you are in contact with a ground facility and have conveyed as much of the above information as time and pertinence permits, your responsibilities from this point are these:

- Comply with advice and instructions, if at all possible.

- Cooperate.

- Ask questions or clarify instructions not understood or with which you cannot comply.

- Assist the ground facility in controlling communications on the frequency. Silence interfering stations.

- Do not change frequencies or change to another ground facility unless absolutely necessary.

- If you do change frequencies, always advise the ground facility of the new frequency and station before making the change.

- If two-way communication with the new frequency cannot be established, return immediately to the frequency where communication last existed.

- Remember the emergency four Cs:
    Confess the predicament to any ground station;
    Communicate as much of the distress/urgency message as possible;
    Comply with instructions and advice;
    Climb, if possible, for better radar detection and radio contact.

Emergency Locator Transmitter (ELT)

While not associated with transponder operations, the Emergency Locator Transmitter (ELT) is related to the subject at hand and is an important element in emergency search-and-rescue (SAR) efforts. In essence, the ELT exists solely to assist in locating downed aircraft.

ELTs are required equipment for most general aviation aircraft, per FAR Part 91.207, although that Part also lists certain exceptions. Essentially, an ELT is nothing but a battery-operated transmitter, along with an externally mounted antenna, that transmits a continuous and distinctive audio signal on the 121.5 and 243.0 frequencies when subjected to crash-generated forces. When transmitting, the life of an ELT is 48 hours over a wide range of temperatures.

Depending on its location in the aircraft, some ELTs can be activated by the pilot in flight. In other installations, the ELT is secured elsewhere inside the fuselage and can't be accessed except on the ground. With this remoted installation (usually near the tail of the aircraft), the transmitter is activated by ground impact, or the pilot can do so after a survivable forced landing when ground impact is comparable to a normal landing.

Because of their importance in SAR efforts, ELT batteries are legal for only 50 percent of their manufacturer-established shelf lives, after which they must be replaced. In the interim, periodic ground checks should be made to determine the battery's viability. These checks, though, can be made only on the ground and in accordance with FAR regulations or the "Emergency Procedures" chapter in *AIM*.

One more note: The FAA urges all pilots to monitor 121.5 while flying around. You might pick up an ELT signal that was beyond the range of an FSS or ATC facility and possibly, through radio contact with a nearby facility, assist in the SAR operation. So, become familiar with what an ELT transmission sounds like and be prepared, if you can, to help a fellow pilot in distress.

Conclusion

I've gone into the matter of emergencies in detail here because of the roles the transponder and radio transmissions play when trouble brews aloft. As rare as troubles are, especially those of a mechanical nature, each of us who flies should be prepared for the worst. That's is one reason simulated forced landings, stalls, short and soft field landings, and so on, are so much a part of initial and recurrent training.

A valuable question to include in any preflight planning is, "What could go wrong?" Then follow that question with another: "If what could go wrong did go wrong, what will I do?" In business parlance, it's called "PPA"—"potential problem analysis." Not only is the "what could go wrong" analyzed, but so are the possible solutions or plans of action if the potential became reality. Those who know what to do, who have planned and practiced, are those who have conditioned themselves to confront airborne problems with reasonable calmness and confidence. Thorough preparation is about 90 percent of emergency survival.

A FEW REMINDERS AND TIPS

As a brief summary, a few reminders and tips about the transponder are perhaps in order:

- Remember to put the transponder in the SBY position after engine start. Change it to ON or ALT only after being cleared for takeoff or during the takeoff roll. Leave it on throughout the flight (mandatory), unless directed otherwise by an ATC facility. On landing, turn it OFF either on touchdown or when you've cleared the runway. Don't keep it in the ON or ALT position while you're taxiing in or out—whether the airport is controlled or not.

- When told by a controller to change from one code to another, jot down the new code on a piece of knee pad paper and then repeat the code back to the controller:

ATC: *Cherokee One Four Six One Tango, squawk five three four zero and ident.*

You: Roger, five three four zero. Six One Tango.

Now push the "Ident" button.

The reason for the readback: It's easy to confuse or transpose digits, as 5-4-3-0, 4-5-3-0, or any other combination. The readback catches any discrepancy. Also, writing down the new code will help you get it straight. It's kind of embarrassing to have to go back to the controller and sheepishly ask, "What was that squawk you wanted?"

- If you have a transponder, with or without Mode C, make it known to ATC in your initial contact:

    Turner Tower, Cherokee One Four Six One Tango over Cordele at three thousand five hundred squawking one two zero zero with Information Delta.

Nowadays, controllers shouldn't have to ask, "Are you transponder equipped?," but they still do, and you still hear it.

- When changing codes, always avoid even a momentary display of 7500, 7600, or 7700. Cycle your changes so that those emergency or radio-loss codes never appear during normal flight operations.

- Become familiar with the transponder radio terminology, as "Squawk VFR," "Squawk Standby," and so on. Again, it's embarrassing to have to ask the controller what he wants you to do.

- Know the emergency transponder and radio procedures by heart. You'll probably never need them, but, like a good Scout, "Be prepared."

CONCLUSION

The transponder is an important element in air traffic control and the ever-increasing drive for safety, and becomes even more essential as new FAA regulations go into effect.

The nice thing about a transponder, though, is that it doesn't require a lot of pilot expertise. In some respects, it's a little like ATIS: It does more for you than you have to do yourself. With ATIS, you just sit and listen. With the transponder, you turn it on, understand the limited phraseology, do as you're asked, and that's it. It—not you—keeps the people on the ground informed of where you are, and, if Mode C-equipped, your altitude. Functioning as it does, it reduces the need for voice communications and contributes to the safety of all of us.

But, a final word is essential. The transponder and what it does should never be allowed to lull you into complacency. Despite the controller's skill and the sophistication of his electronic equipment, it's still the pilot's job to see and avoid. IFR aircraft in instrument meteorological conditions within controlled airspace are assured of horizontal and vertical separation from all other aircraft. As a VFR pilot, however, the most you'll usually get are advisories of the positions of other aircraft and safety alerts when known conflicts seem imminent. In a TCA the control is greater. Otherwise, it's up to you, with whatever help ATC can give you, to do what you always should be doing—protecting your own skin. The final responsibility sits in the left seat of every airplane.

8
Airspace Reclassification

Before getting into the next chapters on the ATC facilities—the Control Tower, Terminal Control Areas, Airport Radar Service Areas, and the rest, a few words about a recently finalized FAA ruling are in order. The reason: The ruling, to one extent or another, affects the nation's airspaces and the very facilities I'll be discussing.

In the process of that discussion, as I explain later, I'll integrate the new elements and terminology of the reclassification with the airspace system of today, to the extent practical. You must recognize, though, that as this book is designed primarily for VFR pilots, many elements that would affect only IFR flight are omitted. Furthermore, what follows does not include all of the reclassification details. It is simply a summary of the primary points of interest the book addresses.

THE FAA AIRSPACE RECLASSIFICATION RULING

Termed "Airspace Reclassification," the ruling, which becomes effective September 16, 1993, is based on the NPRM (Notice of Proposed Rule-Making) 89-28. The final ruling, published in the December 17, 1991, *Federal Register*, fundamentally revises the nomenclature associated with the enroute and terminal airspaces, plus certain related operational changes.

The Ruling—In Summary

Without concern here for its history or the public comments received in response to the NPRM, the objectives of the ruling are:

1. To conform to the international system of airspace designation
2. To simplify U.S. airspace designations
3. To increase standardization of equipment requirements for operations within the various airspaces
4. To revise and/or clarify pilot certificate requirements, VFR visibility and cloud-separation distance, and air traffic assistance offered in each class of airspace
5. To satisfy the responsibilities of the United States as a member of the International Civil Aviation Organization (ICAO)

It's no secret that the United States has been out of step with the international aviation community and the 160-odd nation members of ICAO. Everywhere but here the enroute and terminal airspaces are designated by alphabetical letters, ranging from A through G. The United States, on the other hand, has relied on names or descriptive terms, such as:

Positive Control Area (PCA)
Terminal Control Area (TCA)
Airport Radar Service Area (ARSA)
Air Traffic Area (ATA)
Control Zone (CZ)
Controlled Airspace
Uncontrolled Airspace

Under the new ruling, those designations will disappear in favor of the ICAO lettering system. Consequently, and as also reflected in FIG. 8-1, the reclassification effective September 16, 1993, follows:

Current (US)	*New(ICAO)*
Positive Control Area	Class A Airspace
Terminal Control Area	Class B Airspace
Airport Radar Service Area	Class C Airspace
Airport Traffic Area and Control Zone	Class D Airspace
General Controlled Airspace	Class E Airspace
(Federal Airways, Continental Control Areas, Control Zones without Towers, and a few others)	
Uncontrolled Airspace	Class G Airspace

(Class F Airspace is omitted because the United States has no equivalent to that ICAO classification.)

New FAA Airspace Classifications

Current airspace classification

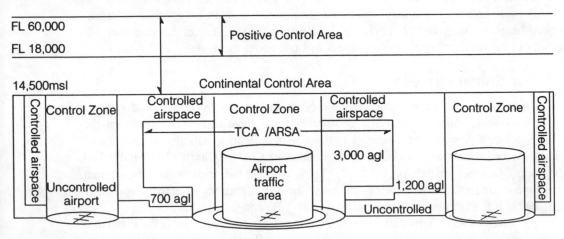

New airspace classification

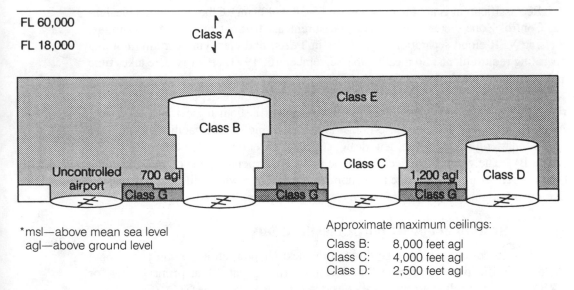

*msl—above mean sea level
agl—above ground level

Approximate maximum ceilings:

Class	
Class B:	8,000 feet agl
Class C:	4,000 feet agl
Class D:	2,500 feet agl

Fig. 8-1. *The more simplified airspace classifications as compared with those current until September 16, 1993.*

The FAA ruling means that, come September 1993, we will no longer have TCAs, ARSAs, ATAs, and the rest. They'll still be there, with a few modifications, but they'll have different designations. The pilot, instead of hearing ". . . cleared into the Dallas TCA," will hear ". . . cleared into the Dallas Class B Airspace."

A more complete summary of the classifications is in TABLE 8-1, which is reproduced from the December 17, 1991, *Federal Register*. To supplement that summary, I'll point out the pertinent operational changes as I talk about the various ATC facilities.

The Ruling Phase-In

To prepare for the reclassification and other changes, aeronautical charts, manuals, textbooks, training materials, and the like will have to be revised. From the pilot's point of view, the first indication of the reclassification will appear on the sectional Aeronautical Charts (SAC) and Terminal Area Control Charts (TAC) published beginning October 15, 1992 (TABLE 8-2). These editions will include both the current airspace classifications and, in parentheses, the classifications effective September 16, 1993. For example, the Sectional will read: "Boston TCA (Class B)."

Although the 37 charts that cover the 48 conterminous states are revised twice a year, the updated versions don't all have the same "effective" and "obsolete" dates. Instead, publication is staggered over a six-month period. Consequently, the series of editions reflecting that first October 15, 1992, change won't be completed until the charts issued by March 4, 1993, appear.

Beginning with the next revision period, September 15, 1993, the legends will be reversed, as "Boston Class B (TCA)." Finally, the dual legends will be eliminated altogether on the charts published between March 3 and August 17, 1994.

Reclassification is not the only change 89-28 will bring. Others include redesigning Control Zones, converting lateral measurements from statute to nautical miles, revising VFR cloud separation regulations in TCAs, and clarifying certain pilot and operating requirements. So meeting the September 16, 1993, effective date takes time and training—thus the planned phase-in.

Although the 4th edition of this book appears well before September 1993, referring to the changes as I move into the next chapters is both logical and necessary. Accordingly, I, too, use the phase-in method by referring to ATC facilities as they are stated today, followed by the new designations in parentheses, for example, "TCA (Class B)." The word "Airspace" might be included, depending on context. Including both references might make the transition from what-is to what-will-be a little easier. Even then, wanna bet there'll be more changes? Better stay tuned.

Clarification about Airport Classifications

To avoid confusion as you go through the next chapter, an important point about airport classifications must be made. The FAA ruling states that primary TCA or ARSA tower-controlled airports are considered to be part of the TCA (Class B) or ARSA (Class C) airspaces. On the other hand, an airport not associated with a TCA

Table 8-1. Airspace reclassifications

Airspace features	Class A airspace	Class B airspace	Class C airspace	Class D airspace	Class E airspace	Class G airspace
Current airspace equivalent	Positive Control Areas	Terminal Control Areas	Airport Radar Service Areas	Airport Traffic Areas and Control Zones	General controlled airspace	Uncontrolled airspace
Operations permitted	IFR	IFR and VFR	IFR and VFR	IFR and VFR	IFR and VFR	IFR and VFR
Entry prerequisites	ATC clearance	ATC clearance	ATC clearance for IFR radio contact for all	ATC clearance for IFR radio contact for all	ATC clearance for IFR radio contact for all IFR	None
Minimum pilot qualifications	Instrument rating	Private or student certificate	Student certificate	Student certificate	Student certificate	Student certificate
Two-way radio communications	Yes	Yes	Yes	Yes	Yes for IFR operations	No
VFR minimum visibility	NA	3 statute miles	3 statute miles	3 statute miles	*3 statute miles	**1 statute mile
VFR minimum distance from clouds	NA	Clear of clouds	500 feet below, 1,000 feet above, and 2,000 feet horizontal	500 feet below, 1,000 feet above, and 2,000 feet horizontal	*500 feet below, 1,000 feet above, and 2,000 feet horizontal	**500 feet below, 1,000 feet above, and 2,000 feet horizontal
Aircraft separation	All	All	IFR, SVFR, and runway operations	IFR, SVFR, and runway operations	IFR, SVFR	None
Conflict resolution	NA	NA	Between IFR and VFR operations	No	No	No
Traffic advisories	NA	NA	Yes	Workload permitting	Workload permitting	Workload permitting
Safety advisories	Yes	Yes	Yes	Yes	Yes	Yes

*Different visibility minima and distance from cloud requirements exist for operations above 10,000 feet msl.

**Different visibility minima and distance from cloud requirements exist for night operations, operations above 10,000 feet msl, and operations below 1,200 feet agl.

Federal Register

Table 8-2. Airspace reclassification transition

Tentative date	Event
October 15, 1992	First sectional aeronautical charts (SAC), world aeronautical charts (WAC), and terminal aeronautical charts (TAC) are published with legends that indicate both existing and future airspace classifications.
March 4, 1993	Initial charting changes are completed for the SAC and TAC.
June 24, 1993	North Pacific, Gulf of Mexico, and Caribbean planning charts are published with legends that indicate both existing and future airspace classifications.
August 19, 1993	Flight Case Planning and North Atlantic Route charts are published with legends that indicate existing and future airspace classifications.
September 16, 1993	New airspace classifications become effective. All charts begin publication with legends that indicate both the new airspace classification and the former airspace classification. All related publications are updated.
March 3, 1994	First charts are published with legends that only indicate the new airspace classifications.
August 17, 1994	All charts are published with legends that only indicate the new airspace classifications.

or an ARSA but that has an operating control tower, is classified as Class D airspace. If a Class D airport has a part-time operating tower, the airport reverts to a General Control (Class E) airspace when the tower is closed.

I need to make this distinction now because, for the sake of simplicity, chapter 9 focuses on the Class D-type of airspace. There is no TCA, no ARSA—simply the tower and the immediate airspace surrounding it.

So, with that brief overview of reclassification, let's get back to the subject of the airport tower facility and its responsibilities.

9
Control Tower, Airport Traffic Areas, and Control Zones

Ground Control has cleared you to taxi to (not on) the active runway. You've completed the pretakeoff check and have moved from the runup area to the runway hold line. With the aircraft at a full stop, you change to the tower frequency and are ready to request takeoff clearance.

About to come into play at this point are two basic elements of the airspace system: the *Airport Traffic Area* (ATA; Class D airspace), and the *Control Zone* (CZ; Class D airspace). In both areas, the tower, which is part of the ATA (Class D), is the primary controlling traffic agency of all flight operations on or in the immediate airport vicinity.

Note: I ask you again to remember that I'm using both the current and the future airspace designations from here on. "Airport Traffic Area," or "ATA," is the correct term until September 16, 1993. That term will then cease to exist and be replaced by "Class D airspace." The same applies to all airspace references in the succeeding chapters.

Although it might bore the knowledgeable, I should begin by clarifying the elements that make up the airport environment: first the Control Tower, then the ATA (Class D airspace), followed by Control Zones. After that, I'll cite examples of the specific FAA-approved radio procedures when communicating in the ATA (Class D) airspace.

WHAT DOES THE CONTROL TOWER DO?

Perhaps the question is so basic that the answers are obvious. On the other hand, perhaps not. Airspace violations and messed-up radio talk raise questions about how well some pilots understand the tower's responsibilities. So does the reluctance of many pilots to venture into a tower-controlled airport if they've been trained at small uncontrolled fields. Anyway, necessary or not, let's pursue the subject.

Simply said, the tower controls all traffic in the *Airport Traffic Area* (Class D airspace), including ground movements on the airport, as cited in chapter 6. In other words:

- To land at or take off from a tower-controlled airport, you must first establish radio contact with the tower controller, obtain specific clearance to take off or land, and then maintain radio contact while within the ATA (Class D airspace).

- To fly through any portion of the ATA (Class D airspace), you must advise the tower controller of your intentions, obtain his clearance to enter the airspace, and thereafter follow his instructions or vectoring.

- If the controller gives you instructions that would cause you to violate a VFR regulation or place you in jeopardy, you must advise the tower of the potential problem and the evasive actions you are taking or intend to take.

Just remember that the tower is not an advisory service, as is unicom or a Flight Service Station. It's a *controlling facility*, which means that its directions must be followed, except in a bona fide emergency or in the situation cited immediately above.

In effect, the controller in the tower is the airport policeman. It is his duty to maintain order in the Airport Traffic Area and to ensure the safe, efficient flow of traffic into, out of, and within that airspace.

THE AIRPORT TRAFFIC AREA

While the following discussion of the Airport Traffic Area pertains primarily to Class D tower-controlled airports (those that are not part of a TCA [Class B] or an ARSA [Class C] airspace), the principles apply equally to all tower airports, regardless of size or complexity. Because of this commonality, I'll omit references to the classes, unless a certain feature is distinctive of one but not the others.

The first commonality is that you won't find the designation of an ATA on any sectional chart. The only clue that such an airspace exists is the blue airport symbol that identifies those airports that have a control tower on the field. Wherever there is an operating tower, there is an ATA. If the tower happens to be part-time, the ATA (in this case, a Class D), becomes a Class E airspace during the hours the tower is closed.

Other Features of an ATA

- The ATA represents the airspace within which the tower is responsible for the flow and control of all traffic—ground and airborne.

- The current lateral limits of the ATA are a circle with a 5-statute-mile radius (4.3 nautical miles), measured outward from the airport center. Under NPRM 89-28, the ATA dimensions will still be 5 statute miles, but all references to the radius will be stated in nautical miles: (4.3nm = 5.0 sm).

- The current vertical limit of an ATA is 3,000 feet agl. That will be reduced to 2,500 feet agl under reclassification 89-28, but the altitudes can vary upward somewhat with individual airports. Above the vertical limit, the ATA no longer exists. Accordingly, when passing over an ATA (Class D) above its published limit, no radio contact with the tower is necessary or wanted. (TCA [Class B] and ARSA [Class C] airspaces have different radio requirements, as will be discussed later.)

- As an ATA exists only when there is an operating control tower, the radio requirements outlined earlier relative to the tower apply: Permission to enter, leave, or transit any portion of the ATA is mandatory.

- If the tower is part-time, there is no ATA when the tower is closed, and the airport is considered uncontrolled. Under reclassification 89-28, the Class D airspace then becomes a General Controlled (Class E) airspace. With a closed tower, both now and when reclassification becomes effective, the routine position reports discussed in chapters 2, 3, and 4 are made via the tower frequency.

CONTROL ZONES (CLASSES B, C, AND D AIRSPACES)

The sole purpose of a Control Zone (CZ) is to provide proper separation of arriving and departing IFR aircraft in IMC (Instrument Meteorological Conditions). When the weather is VFR in the airport environment, the CZ doesn't exist. It is still on the sectional, but you don't have to contend with it as an airspace as a VFR pilot.

Like ATAs, CZs are associated with all tower-operated airports and some non-tower fields. In the latter case, the CZ is classified as Class E airspace.

Whether the airport has a tower or not, and unlike the ATA, the CZ and its lateral dimensions are clearly defined on sectional charts. If you look at the Pendleton, Oregon, example in FIG. 9-1, you'll note the broken blue lines that form a circle around the airport, plus the slight extension of the lines to the west. Further, if you measure the radius of the circle from the airport, you'll see that it's 5 sm (4.3 nm) from the airport center. These broken lines, plus the extension, establish the lateral dimensions of the CZ.

At the same time, the circular portion defines the limits of the ATA. Any extension, as in the Pendleton example, however, exceeds the 4.3-nm limit and is thus part of the CZ but not the ATA. So while the ATA itself is not identified on the sectional, the boundaries of the CZ, excluding any extensions, do establish the ATA limits and its area of responsibility.

Another advantage of these chart markings: When arriving or transmitting in an ATA (Class D airspace), plan to make your radio call to the tower about 10 miles out

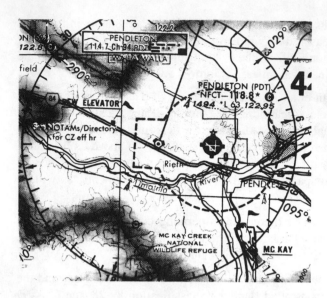

Fig. 9-1. *Pendleton, Oregon, has a tower, thus an ATA and a Control Zone, as depicted by the broken keyhole-shaped design. The ATA, however, is never shown on any chart.* NFCT *means "Nonfederal Control Tower."*

from the broken circle. That gives you time to establish contact, report your position, and receive clearance into the ATA (Class D) before actually penetrating the area.

Some Points to Keep in Mind about Control Zones

- Although the lateral limits of a CZ generally coincide with those of an ATA, its vertical limits rise from the surface to 14,500 feet msl, which is the base of the Continental Control Area (FIG. 9-2).

- When the airport is above VFR minimums, radio communication with the tower is not required to fly through any portion of the CZ that is outside the lateral or vertical limits of the ATA (Class D) airspace.

    To clarify, refer again to FIG. 9-2. When you're in the shaded area, you're in the CZ but not the ATA. Thus in VFR conditions, no radio contact in a Class D ATA is necessary because, in effect, there is no CZ.

- At an airport with a part-time tower, the CZ continues to exist even when the tower is closed. That's because an Approach Control or a Center can provide approach or departure radar service for IFR aircraft in Instrument Meteorological Conditions. Both the airport and the CZ, however, become a Class E airspace when the tower is not in operation.

- When the airport is below basic VFR weather minimums, a Special VFR (SVFR) clearance through either the tower or a Flight Service Station is required prior to operating in a CZ.

- Some of the busier airports, particularly TCAs (Class B airspace), prohibit SVFR clearances. Those airports are identified on the sectional by a series of Ts that form the broken circle of the CZ.

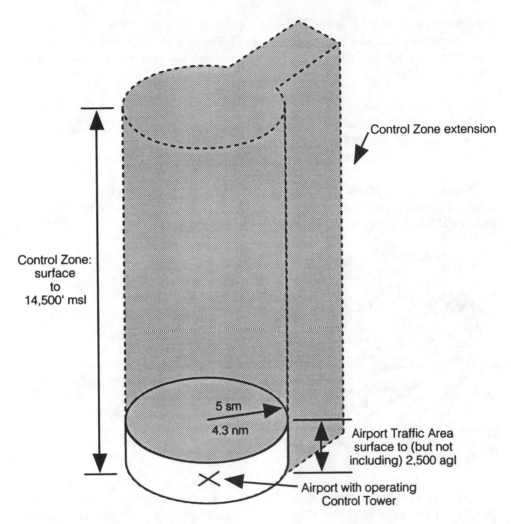

Control Zone extension

Control Zone:
surface
to
14,500' msl

5 sm

4.3 nm

Airport Traffic Area
surface to (but not
including) 2,500 agl

Airport with operating
Control Tower

Fig. 9-2. *The basic structure of a Control Zone. Radio communications with the tower, however, are not necessary within the shaded area in VFR weather conditions.*

• Effective September 16, 1993, the term *Control Zone* will no longer be used. Instead, the CZ will be considered a part of the airspace with which it is associated—Class B, C, or D airspaces, except when a Class D tower is closed. Then, of course, the CZ becomes a Class E.

Airports with Control Zones (Class E airspace) but No ATA

If you check any sectional, you'll find lots of airports colored in magenta, signifying no control tower and thus no ATA. Around some of these fields, though, are the broken lines identifying Control Zones. A logical question is: "How come? What is there about them that justifies a Control Zone?" The answer: CZ exists when one or a combination of the following is located on the field and can provide the essential weather services:

- An operating tower and a qualified weather observer
- An operating Flight Service Station
- A National Weather Service Office
- A qualified weather observer

One example is Muscle Shoals, Alabama (FIG. 9-3). Even though it has no tower, it has a CZ because a Flight Service Station is on the field, as indicated by the heavy-lined rectangle and the "FSS" in magenta above the airport name.

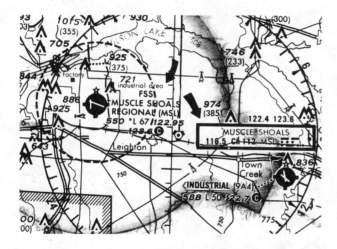

Fig. 9-3. *Muscle Shoals, Alabama, is one example of a nontower airport with a Control Zone and an FSS on the field.*

Another example is FIG. 9-4, the Goodland Airport in northwest Kansas. Here you have a no-tower field, the Flight Service Station serving the area is some 260 miles away in Wichita, and the chances of a National Weather Service Office being located on the field are remote. Consequently, it must have a qualified weather observer.

Another question might come to mind at this point: If a CZ exists to expedite the safe arrivals and departures of IFR aircraft in instrument weather conditions, and the airport has no tower and is miles from any ATC facility, how, then, are IFR aircraft controlled?

The answer: Taking the case of Goodland, IFR aircraft landing or taking off from Goodland in IMC weather are controlled by the Denver Air Route Traffic Control Center through a remote microwave link back to the Center that permits radar contact and vectoring down to the ground.

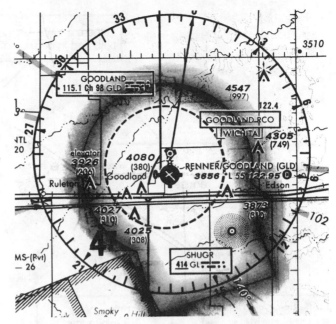

Fig. 9-4. *Goodland has a qualified weather observer, which justifies a Control Zone.*

On the other hand, if an airport is beyond the remoted radar range and contact with the target is not possible below a certain altitude, the Center controller uses non-radar procedures to ensure the proper separation of the IFR aircraft. In these cases, however, the airport would not have a control zone. Nor would it have a CZ if its traffic volume made installation of a remoted outlet impractical.

If the conditions are VFR at a Goodland, you are not required to contact anybody before landing or taking off, despite the obvious safety precaution of keeping others informed of your presence and intentions via the CTAF (unicom or 123.6).

On the other hand, as soon as conditions drop below basic VFR minimums, a Special VFR clearance must be obtained through the responsible Flight Service Station before you can enter the Control Zone or depart from it. (Keep in mind that I'm talking about a nontower airport. If a tower were on the airport, it would be the source to contact for the clearance.)

At airports where the tower, FSS, or weather observer is on duty only part-time, the CZ is also part-time and is effective only during the hours listed in the *Airport/ Facility Directory* and NOTAMs.

At this point, if you are still a bit confused and wondering "What do I do when and with whom? " refer to FIG. 9-5.

Specific examples of radio contacts are suggested later in this chapter. The principal purpose of discussing ATAs and CZs here is to try to clarify the meaning of each and the conditions under which communications with the tower or Flight Service are essential.

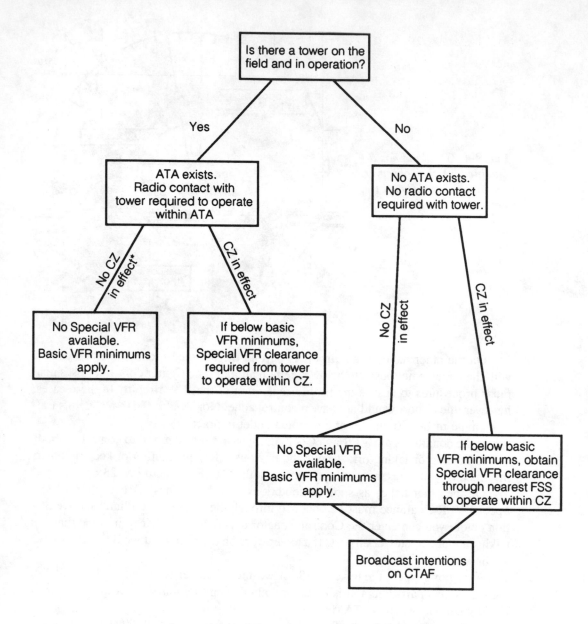

Notes: Special VFR is not available within some high-activity Control Zones.
Control Zone effective hours are listed in the A/FD
*Tower airports without CZs are extremely rare, but do exist.

Fig. 9-5. *This diagram helps determine the communications requirements at airports with ATAs and Control Zones.*

DETERMINING THE TOWER FREQUENCY

For the VFR pilot, two sources are available to determine tower frequencies—the sectional chart and the *Airport/Facility Directory (A/FD)*.

The sectional always publishes the frequencies in two locations: adjacent to the airport name, with "CT" preceding the frequency (FIG. 9-6); and on the reverse side (usually) of the legend flap (FIG. 9-7).

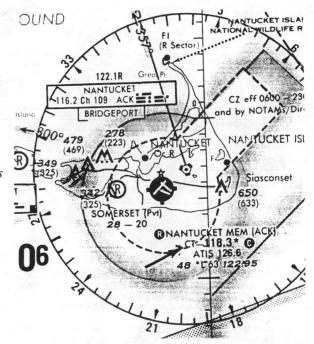

Fig. 9-6. *The easiest place to find the tower frequency is next to the "CT" ("Control Tower") on the sectional.*

The *A/FD* is published six times a year, each issue having approximately a two-month validity period. One advantage of this source is that, for a given location, it gives all the frequencies under the "Communications" heading: Unicom, Flight Service (either on the field or remoted), Approach/Departure Control, ATIS, Ground Control, Clearance Delivery, and tower (FIG. 9-8). The sectional is never this complete.

Just be sure you always use the current editions of the sectional and A/FD. Frequencies do change.

DOING WHAT THE TOWER TELLS YOU

Yes, you are the pilot in command, with certain responsibilities and authority. And, yes, the tower is there to serve you, along with all the other pilots in the area.

CONTROL TOWER FREQUENCIES ON NEW YORK SECTIONAL CHART

Airports which have control towers are indicated on this chart by the letters CT followed by the primary VHF local control frequency. Selected transmitting frequencies for each control tower are tabulated in the adjoining spaces, the low or medium transmitting frequency is listed first followed by a VHF local control frequency, and the primary VHF and UHF military frequencies, when these frequencies are available. An asterisk (*) follows the part-time tower frequency remoted to the collocated full-time FSS for use as Airport Advisory Service (AAS) during hours tower is closed. Hours shown are local time. Ground control frequencies listed are the primary ground control frequencies.

Automatic Terminal Information Service (ATIS) frequencies, shown on the face of the chart are normal primary arrival frequencies. ATIS operational hours may differ from control tower operational hours.

ASR and/or PAR indicates Radar Instrument Approach available.

"Mon-Fri" indicates Monday thru Friday.

CONTROL TOWER	OPERATES	TWR FREQ	GND CON	ATIS	ASR/PAR
MUIR AAF	0800-2400 MON-FRI 0800-1600 SAT-SUN EXCLD HOL	126.2 241.0	139.0 265.6		
NANTUCKET MEM	0600-2300 MAY 15-SEP 30 0600-2100 OCT 1-MAY 14	118.3 257.9	121.7	126.6	
NEWARK INTL	CONTINUOUS	118.3 257.6	121.8	ARR 115.7 ARR 134.825 SOUTH DEP 132.45	
NEW BEDFORD	0630-2200	118.1 239.0	121.9	126.85	
NORTHEAST PHILADELPHIA	0600-2300	126.9 349.0	121.7	121.15	
NORWOOD MEM	0700-2200	126.0	121.8	119.95	
ONEIDA CO	CONTINUOUS MON-FRI 0600-2400 SAT-SUN	118.1* 291.7	121.9 241.0	118.7	
OTIS ANGB NF	CONTINUOUS	121.0 294.7	121.6 275.8		ASR
PEASE ANGB NF	CONTINUOUS	128.4 255.9	120.95 275.8	273.5	ASR
PORTLAND INTL	0600-2400	120.9 257.8	121.9	119.05	ASR

Fig. 9-7. *Another place to find the tower frequency: inside the legend flap of the sectional.*

NANTUCKET MEM (ACK) 3 SE UTC-5(-4DT) 41°15'11"N 70°03'39"W   **NEW YORK**
48   B   S4   FUEL 100LL   ARFF Index A   **H-3J, L-25D**
RWY 06-24: H6303X150 (ASPH) S-75, D-170, DT-280 HIRL 0.3%up NE.   **IAP**
  RWY 06: MALSF. VASI(V4L)—GA 3.0°. Thld dsplcd 539'.   **RWY 24:** ALSF1.
RWY 15-33: H3999X150 (ASPH) S-60, D-85, DT-155 MIRL
  RWY 15: REIL.   **RWY 33:** REIL. VASI(V4R)—GA 3.0°TCH 43'.
RWY 12-30: H3125X50 (ASPH) S-12
AIRPORT REMARKS: Attended Oct 15-May 15 1100-0200Z‡ and May 16-Oct 14 1100-0400Z‡. For fuel after hours call 508-228-5159 beeper 326 or 228-5125. Be aware of hi-speed military jet and heavy helicopter tfc vicinity of Otis ANGB. Rwy 12-30 VFR/Day use only aircraft under 12,500 lbs. Arpt has noise abatement procedures ctc arpt manager 508-325-5300. When twr clsd ACTIVATE MALSF Rwy 06; HIRL Rwy 06-24; MIRL Rwy 15-33 and all Twy lgts—CTAF. Rwy 24 ALSF-1 unmonitored when arpt unattended. Control Zone effective 1100-0400Z‡.
  NOTE: See SPECIAL NOTICE—Simultaneous Operations on Intersecting Runways.
WEATHER DATA SOURCES : LAWRS.
COMMUNICATIONS: CTAF 118.3
  ATIS 126.6 (508-228-5375) (1100-0200Z‡) Oct 1–May 14, (1100–0400Z‡) May 15–Sept 30.
  UNICOM 122.95
  BRIDGEPORT FSS (BDR) TF 1–800–WX–BRIEF. NOTAM FILE ACK
  RCO 122.1R 116.2T (BRIDGEPORT FSS)
  Ⓡ**CAPE APP/DEP CON** 126.1 (1100-0400Z‡) May 15–Sept 30, (1100–0300Z‡) Oct 1–May 14.
  BOSTON CENTER APP/DEP CON 132.9 (0400-1100Z‡) May 15–Sept 30, (0300–1100Z‡) Oct 1–May 14.
  TOWER 118.3 May 15–Sep 30 (1100–0400Z‡), (1100–0200Z‡) Oct 1–May 14.
  GND CON 121.7 **CLNC DEL** 128.25
RADIO AIDS TO NAVIGATION: NOTAM FILE ACK. VHF/DF ctc NANTUCKET TOWER
  (H) VORTAC 116.2 ACK Chan 109 41°16'54"N 70°01'38"W 236°2.3 NM to fld. 100/15W. Enroute ATIS broadcast by New York ARTCC 1200-0400Z‡ consists of significant meteorlogical information and delays at the Metropolitan New York arpts, may also be obtained by dialing 718-995-5049.
  WAIVS NDB (LOM) 248 AC 41°18'40"N 69°59'14"W 239° 4.8 NM to fld.
  NDB (HH–ABW) 194 ■ TUK 41°16' 07"N 70°10' 50"W 115° 5.5 NM to fld.
  ILS/DME 109.1 I-ACK Chan 28 Rwy 24 LOM WAIVS NDB. ILS unmonitored when twr clsd.

Fig. 9-8. *The third reference source for the tower frequency is the* A/FD.

Neither condition, however, alters the fact that no pilot has the right to go his way regardless of the Tower's instructions. To repeat what I've already said, unless an emergency suddenly develops or adherence to an instruction would violate a FAR, you must comply with the tower's directives.

The controller is spacing, separating, coordinating, and overseeing all the traffic within his area of responsibility. He has a plan to get everyone up or down with minimum delays. He can't have the whole operational pattern disrupted because some character does a 360 on the downwind leg or chooses to land on Runway 21 when Runway 18 is the active runway. Perhaps judgment decrees that a 360 is essential for spacing or safety. If so, advise the tower before starting the maneuver. Perhaps Runway 21 is better because of the wind. Fine, but get permission before you switch to another pattern.

Does any of this sound fundamental? Sure, but such unannounced or unapproved deviations from instructions are not rare. Stick around a controlled airport for a while. Listen and observe. You'll see.

WHAT IF YOU DON'T UNDERSTAND?

Some controllers speak rapidly; some occasionally slur their words; some might use a term or issue an instruction you don't understand; perhaps your radio reception is fuzzy; maybe someone cuts in just as the tower tells you what to do and all you hear are squeals and squawks. Whatever the case, you don't understand the controller's message.

Here is where uncertainty can have unpleasant consequences. If you don't understand, don't just "Roger" the instruction and hope that whatever you do is the right thing.

One of the basic elements of communications is understanding. The word *communication* comes from the Latin *communicare*, meaning "to share; to make common." When the words that flow between speaker and receiver are understood—when the receiver has the same idea in his mind that the speaker had in his—there is a sharing, a commonality. Hence, there is *communication*.

The reasons for breakdowns in human relations, in business transactions, and in international relations are myriad. Perhaps the one factor most responsible is the failure of people—or nations—to communicate in ways that nurture mutual understanding. It's too easy to assume that what I say has the same meaning to you that it has for me. Messages are distorted by wishful hearing, wandering attention, mistrust, and word choice (i.e., jargon, slang, unfamiliar terms, and imprecise words).

To an Englishman, a bonnet is the hood of his car, while Americans typically think of baby hats or the Fifth Avenue Easter Parade. A trolley is a restaurant serving cart in Britain, but Americans conjure up thoughts of San Francisco's cable cars. The tower tells you to "fly the final"—perfectly clear words, and you say to yourself, "What the hell! I'm flying, and I'm on the final. What's he talking about?" He means to extend the final by making S-turns for better spacing. Communication is lost in the haze of jargon.

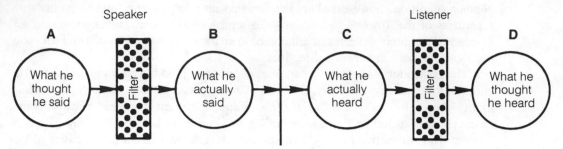

Fig. 9-9. *The ''filters'' are what so frequently distort speaker-to-listener communications.*

Figure 9-9 illustrates part of the ongoing communication problem: It's a long way from A to D, and the road can be fraught with many unintended detours.

Compounding the inherent causes of poor communication is the human tendency to try not to appear stupid. Most people are reluctant to admit they didn't understand a directive or an instruction, especially if they feel the communicator expects instant comprehension. It's a matter of preserving one's self-esteem, of not losing face. One of the more meaningless questions an instructor, a boss, or a parent can ask is "Do you understand?" Unless complete trust between the two parties exists, not many people are willing to say, "I didn't understand a thing you said." That's admitting ignorance or—worse—stupidity.

Pilot and controller talk to each other on a one-to-one basis. However, an unseen listening audience out there captures everything the two are saying. Well aware of this, the pilot who is uncertain about radio communications faces the added concern of exposing his uncertainty to his airborne compatriots. So he "Rogers" the instructions and prays that everything works out all right.

My point here is to air the concerns every pilot has—or had—about radio procedures and communications. If you've got a few thousand hours in your log, the concerns have (I hope) evaporated. The less experienced you are, the greater the concerns are likely to be—which, of course, is one reason why those who learned to fly at uncontrolled airports venture most hesitantly into the larger aerodromes.

As I emphasized in an earlier chapter, know what you're supposed to say, and then practice it—even at home with a tape recorder. Practice sharpens your "sending" ability. Then, while in the air, listen long and hard to what the tower tells other pilots. If you've never used Approach or Center, tune in to the appropriate frequency and do some eavesdropping. Just getting the feel of the verbal exchanges helps.

If, despite all your practice, you still don't understand what the tower tells you, ask for immediate clarification:

"Tower, say again."
"Tower, Cherokee Six One Tango. Say again more slowly."
"Your transmission was garbled. Say again, please."
"Am unfamiliar with the term. Please explain."

"You were cut out. Please repeat instructions."

"Am unfamiliar with the area. Please identify location of reporting point." (Tower told you to report "over the twin stacks," but where are the "twin stacks"?)

The controller is similar to a coach, and the pilots on the ground or in the air are the players. The controller is calling the plays. He's in charge. The players are expected to do what he says. If a "play" won't work, the "coach" should be notified. If there is misunderstanding or confusion, it had better be cleared up now, because the "game" in the air is far more consequential than any earthbound contest.

Yes, there is a "team" relationship between pilot and controller. Without clear communications between them, potential problems abound. Working together in harmony, each is a winner.

THE TAKEOFF CONTACT

The start of this chapter had you ready for takeoff. Let's continue from there. Assume that you're number one to go and are still in the runup area, some 75 feet or so from the hold line. After the pretakeoff checks are complete, taxi to the hold line and come to a complete stop. At this point, you're clear to switch from Ground Control to the tower frequency. Before making the call, however, check your volume, and be sure you're ready to transmit as well as receive on the proper frequency.

Why call at the hold line (FIG. 9-10) and not when you have completed your runup? Suppose that an aircraft is on a relatively short base about to turn final. If you're still back in the runup area, you probably won't have enough time to get to the runway before the landing craft is on your tail. Hence, the tower would probably tell you to "taxi closer and hold." But if you're at the hold line, the tower might be able to clear you for an immediate takeoff. Through this simple maneuver, traffic is expedited and aircraft on the ground behind you have less idling and fuel-burning time.

When you're really ready to go, the call is nothing more than:

You: Cedar Tower, Cherokee One Four Six One Tango ready for takeoff, east departure.

Twr: *Cherokee Six One Tango cleared for takeoff. East departure approved.*

You: Roger, cleared for takeoff, Cherokee Six One Tango.

Remember KISS. No need for something like this:

You: Cedar Control Tower, this is Cherokee One Four Six One Tango, Over.

Twr: *Cherokee One Four Six One Tango, Cedar Tower.*

You: Cedar Control Tower, this is Cherokee One Four Six One Tango. We're ready to take off on Runway Three Six. We would like to leave the pattern and depart to the east. Over.

Twr: *Cherokee Six One Tango cleared for takeoff, east departure approved.*

You: Roger. Six One Tango. East departure approved.

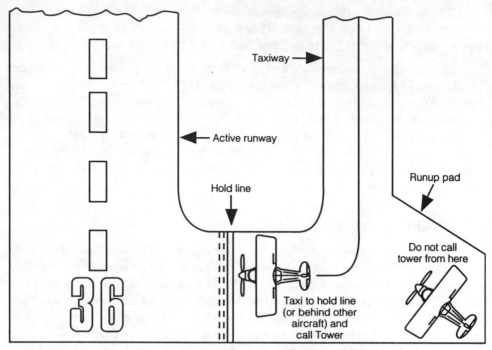

Fig. 9-10. *Call the tower at the hold line—not back in the runup area.*

That transmission consumed 52 words, not counting the tower's replies. What have you said that you didn't say in 21 words in the other departure call? Nothing. And the tower knows you're going to use Runway 36, so why state the obvious?

Now let's assume that you've made the call and the tower acknowledges with: "Cherokee Six One Tango, hold for landing traffic." Your reply should indicate your understanding and compliance: "Six One Tango holding."

The landing aircraft has landed, and the tower contacts you again: "Cherokee Six One Tango, taxi into position and hold." Your reply: "Position and hold, Six One Tango."

You haven't merely "Rogered" the instruction. You've confirmed to the tower that you have understood and will comply with the instructions.

When the landing craft has cleared the runway, the tower calls you once more: "Cherokee Six One Tango, cleared for takeoff." What's missing? Should you take off under the possibly false assumption that your eastbound departure has been approved? No. You might start your run, but don't assume anything. Contact the tower again:

You: Tower, Cherokee Six One Tango. Is eastbound departure approved?

Twr: *Cherokee Six One Tango, Roger, eastbound departure approved. Remain north of the buildings.*

You: Wilco ["Will comply"], Cherokee Six One Tango.

Another situation: after your initial "ready for takeoff" call, the tower says: "Cherokee Six One Tango cleared for immediate takeoff."

What's the controller saying? He's telling you to get out on that runway and *go*. "Immediate" means *immediate*. You don't idle your way to 36 and take your sweet time setting everything up. Taxi out as rapidly and safely as you can to the center of the runway, line up the airplane, and apply power now.

This sort of instruction doesn't require your acknowledgment. The fact that your aircraft is moving is indication enough that you have received the message. If you sit on the runway in takeoff position with no visible activity on your part, you're certain to hear from the person in the elevated enclosure.

A variation of this situation: Responding to your ready-to-go call, the tower comes back with: "Cherokee Six One Tango cleared for immediate takeoff or hold short." It's decision time. If you're really ready, taxi out and get going; the rolling aircraft confirms your intentions. Otherwise, tell the tower what you're going to do. Silence leaves the controller in a state of uncertainty—which makes controllers very unhappy. Don't play *I've Got a Secret*. Quickly reply, "Cherokee Six One Tango will hold."

One other example of these takeoff contacts: Ahead of you at the hold line are two other aircraft. You've completed your runup and are ready to go, but the other planes are just sitting there. If, for example, they're waiting for IFR clearances, they might remain sitting for some time, while you stay patiently in line burning up fuel. So what do you do? The logical thing is to call the tower. If possible, the tower will clear you ahead of those who precede you:

You: Cedar Tower, Cherokee One Four Six One Tango, number three in sequence, ready for takeoff. Request east departure.

Twr: *Cherokee Six One Tango. Taxi around the Cessna and Mooney. Cleared for takeoff. East departure approved.*

You: Roger, cleared for takeoff, Cherokee Six One Tango.

CHANGING FREQUENCIES AFTER TAKEOFF

For a variety of reasons, you might want (or need) to change frequencies shortly after taking off. Perhaps you're going to a nearby airport and want to monitor the traffic; your departure route takes you over another airport and clearance to cross it is required; you want to contact Departure Control; you need to open or modify your VFR flight plan with Flight Service; and so on. Whatever the reason, always ask for and receive approval from the tower to change to another frequency as long as you are still within the Airport Traffic Area—the 5 mile radius. The request is: "Cedar Tower, Cherokee Six One Tango requests frequency change."

In all likelihood, the tower will approve the request. If local traffic is heavy or visibility limited, however, the controller might want you to stay with him until you are well clear of his area. Depending on the circumstances, his response might be: "Cherokee Six One Tango, frequency change approved," or "Cherokee Six One

Tango, remain on this frequency. I'll have traffic for you," or "Cherokee Six One Tango, stay with me until clear of the area."

The point is: Don't leave the tower while within the ATA until the tower has approved the change. The controller might need to contact you for any number of reasons, and you are still within airspace under his control.

TAKEOFF WITH CLOSED PATTERN

This time, you want to sharpen your landings with a few touch-and-gos:

You: Cedar Tower, Cherokee One Four Six One Tango, ready for takeoff. Request closed pattern.

Twr: *Cherokee One Four Six One Tango Roger. Cleared for takeoff. Closed pattern approved.*

From this point on, you usually won't have to initiate any further contacts with the tower. This is a controlled airport, and the tower knows your intentions. All that's necessary is to acknowledge the controller's transmissions to you:

Twr: *Cherokee Six One Tango, cleared for touch-and-go, Runway Three Six.*

You: Roger. Cherokee Six One Tango.

Again, however, keep the tower informed. When you decide that you've had enough practice, it's a good idea on the downwind leg to tell the tower that this will be a full stop—even before the controller clears you for another touch-and-go. Initiating the call early on the downwind might help the controller space other aircraft that are either landing or taking off:

You: Cedar Tower, Cherokee Six One Tango, will be full stop this time.

Twr: *Cherokee Six One Tango, Roger. Cleared to land, Runway Three Six.*

Whether you initiate the call or respond to the controller's clearance for another touch-and-go, be sure to inform him of your intentions. Controllers don't like surprises. However, unlike at multicom or unicom airports, do not call the tower on downwind or when turning base and final, unless there is a special reason to do so. The controller knows what you're doing, and these calls only clog the air.

APPROACHING-THE-AIRPORT CONTACTS

You're on a cross-country flight and are nearing your destination airport. Just to make it simple, let's say that there's no TCA or ARSA associated with the field. Be a good Scout before making the initial contact: Be prepared!

An airline captain told this story of what came over the air one time. As he was approaching a field, he heard a charter carrier talking to the tower. With names and locations changed to protect the guilty, the call went something like this:

Pilot: Mayflower Tower, this is Rocky Charter Three Niner Niner Uniform.

Twr: *Rocky Charter Three Niner Niner Uniform, go ahead.*

Pilot: Tower, Three Niner Niner Uniform, we're over . . . [to the first officer, with an open mike] Where are we, Harry? [Pause] We're over North Centerville for landing at Mayflower.

Twr: *Roger, Three Niner Niner Uniform. Say your altitude.*

Pilot: Altitude is . . . What's our altitude, Harry? [Pause] Altitude is seven thousand five hundred.

Twr: *Roger, Three Niner Niner Uniform. What's your airspeed?*

Pilot: Let's see, airspeed . . . how fast we going, Harry? [Pause] 375 knots, Tower.

Twr: *What are you squawking, Three Niner Niner Uniform?*

Pilot: We're squawking . . . Harry, what are we squawking?

At this point, the tower broke in:

Twr: *Three Niner Niner Uniform, would it be all right if we talked to Harry?*

Not having personally heard this exchange, I can't vouch for its authenticity. The airline captain, however, left no doubt in my mind that it was almost a verbatim copy of what transpired.

Know what you're going to say, and if you're a little uncertain, rehearse it to yourself before you pick up the mike and press the button. Also, if you've set things up properly, you've listened to the ATIS and monitored instructions to other aircraft on the tower frequency. With these transmissions, you know the winds, altimeter setting, active runway, and traffic pattern direction. Now you're equipped to plan your approach because you have a good idea what the tower will tell you when you establish contact.

At what point should you make the initial tower call? To repeat, it must be made before you enter the 5 mile ATA (class D) radius. To tower personnel, you're a stranger, and they don't appreciate strangers who bounce into their territory unannounced. A reasonable rule of thumb is to introduce yourself about 15 miles out or at the little red flags on the sectional that identify the various visual reporting points.

After listening to the last transmission to be sure the air is dead, follow this sequence in the call:

1. Tower name
2. Aircraft type
3. N-number
4. Position
5. Altitude
6. Squawk (if transponder-equipped)
7. ATIS identification

The initial call should go like this:

> Cedar Tower, Cherokee One Four Six One Tango over Grand Lake Dam, level at
> three thousand five hundred, squawking one two zero zero with Information Lima.

The tower's response can vary, but if everything is normal, you'll soon hear the
controller say:

> **Twr:** *Cherokee One Four Six One Tango, enter left base for Runway One Eight.*
> [plus winds and altimeter if there's no ATIS]
>
> **You** Roger, left base for One Eight. Cherokee Six One Tango.

Once again I stress the importance of tersely repeating the tower's instructions. If
you're told to "enter left base for One Eight," confirm your understanding with "left
base for One Eight."

If the instruction is to "cross midfield at two thousand three hundred for left
downwind," something like this ensures that you're both on the same page in the same
hymnbook: "Roger, midfield at two thousand three hundred, left downwind."

You're instructed to "report two miles east of the (airport, buildings, city, bridge,
water tower, power plant, or whatever)." Your reply is: "Roger, report two east (of
whatever). Cherokee Six One Tango."

These responses are more communicative than a "Roger." Theoretically,
"Roger" means that you have understood and will do what you've been told. But do
you understand? Probably, but the tower can't be sure. Of course you don't want to
clutter up the airwaves with needless commentaries, but suppose that part of the tower's transmission was garbled or indistinct and you think the runway is 24 when it is
actually 34. To plan your pattern accordingly could have disruptive effects. This sort
of confusion is unlikely if you monitored ATIS and the tower beforehand, but confusion is still possible.

As just one example, I heard a Cessna 172 not long ago that was piloted by Lieu-
tenant Confusion or Captain Stupid. It was coming into the Kansas City Downtown
Airport from the south, with 01 as the active runway. After I heard the Cessna first on
Approach Control, I switched to the tower to announce my position (I was not using
Approach, just eavesdropping). A few minutes later, the Cessna turned up well north
of the airport and hesitantly contacted the tower. The tower clearly identified the
active as Runway 01, with right-hand traffic. The Cessna dutifully rogered the information. Next, the tower told the Cessna to fly a heading of 190 degrees on the downwind. Again, the Cessna rogered. Apparently, the tower was keeping a close eye on
the errant aircraft, because the next question was: "Cessna Zero Zero Zero Zero, are
you landing on Runway Three?" To which the Cessna rogered again.

Now, Runways 01 and 03 happen to cross each other, and cross-traffic landings
and takeoffs, if not controlled, can lead to messy situations. Fortunately for at least
two aircraft, traffic was sufficiently spaced and the controller sufficiently alert to per-

mit the visiting Cessna, which was now on a tight base, to continue to Runway 03. No harm was done.

Whether Confusion or Stupid was in command doesn't matter. It is unlikely that Approach vectored the pilot over and well north of the airport when a straight-in-approach from the south to Runway 01 was the shortest distance between the two points. The pilot must have confused Runway 03 with Runway 01, and he was flying a downwind at 210 degrees instead of the required 190 degrees. And he calmly rogered the question, "Are you landing on Runway Three?"

If the Cessna pilot had rogered less and repeated at least a couple of the instructions just once, the tower could have straightened things out before a potential hazard had arisen. "Roger" doesn't necessarily ensure commonality of understanding.

Let's continue with the approach and landing. If you're transponder-equipped, you might receive no further instructions from the tower other than "Cherokee Six One Tango, cleared to land." To which you reply "Roger, cleared to land, Cherokee Six One Tango." Roger. You've got it.

On the other hand, and particularly if you have a transponder, the tower might be in frequent communication to advise you of other aircraft in your vicinity:

Twr: *Cherokee Six One Tango, traffic is a Cessna at one o'clock, two miles, westbound at two thousand.*

[Pause while looking]

You: Negative contact, Cherokee Six One Tango.

Or, if you spot the traffic: "Cherokee Six One Tango has the Cessna."

Another situation: you're advised of the Cessna at one o'clock. You don't see it, and so inform the tower. A minute or so later, you spot it at your two o'clock position. At this point, tell the tower—even though the traffic is well to your right and presents no possible hazard: "Cedar Tower, Cherokee Six One Tango has the Cessna." The tower will acknowledge your message, often with a "thank you."

In a similar scenario, you hear the tower call another aircraft in your general area, alerting the pilot to your presence: "Cessna Eight Niner Golf, traffic is a Cherokee at ten o'clock, two miles, also westbound. Altitude unknown."

You have a reasonably good idea that that's you, so get on the air and help everybody by reporting your altitude: "Cedar Tower, Cherokee Six One Tango is at two thousand eight hundred, descending to one thousand seven hundred (pattern altitude)." You might not get an acknowledgment, but that's beside the point. You've kept the other parties informed—at least the interested parties.

This is as good a time as any to mention that when you're given traffic in a "clock" position (nine o'clock, one o'clock, and so forth), the position is based on your *ground track*, as shown on radar. The traffic's position might be different from your point of view, because of your wind correction angle. For example, you're tracking over the ground at 270 degrees, but because of a northerly wind, your heading is

300 degrees. If you are advised that you have a "target" at "12 o'clock," that means that the other aircraft is on your 270 degree track, not straight ahead of the nose of your airplane at 300 degrees. The target then, is actually at about 11 o'clock in relation to the direction in which your airplane is pointed. Keep this in mind, especially under strong wind conditions when you need to crab to maintain your desired course.

Now let's continue and say you're nearing the airport. Once again, the tower calls you:

Twr: *Cherokee Six One Tango, you're number two to land behind the Mooney on downwind.*

You: Roger, number two to land. Negative contact on the Mooney [or "and we have the Mooney"]. Cherokee Six One Tango.

When the spacing is proper between you and the Mooney, or the Mooney is about to touch down, the tower will give you final clearance:

Twr: *Cherokee Six One Tango, cleared to land, Runway Three Six.*

You: Roger, cleared to land Three Six, Cherokee Six One Tango.

Once you're on the ground, be sure to stay on the tower frequency until you have turned off the active runway and the tower has cleared you to the ramp or advised you to contact Ground Control. You have no way of knowing what might be happening behind you that would require the tower to issue you an emergency instruction. The call from the tower will probably be brief—no more than "Cherokee Six One Tango, contact Ground point niner." Remember that the Ground Control frequency is typically 121.7, 121.8, or 121.9. Consequently, the first three digits are often dropped.

Occasionally, especially if the tower is busy, the controller might fail to tell you to contact Ground, even though you are clear of the runway. In such cases, go past the taxiway hold line, come to a complete stop, and call the tower:

You: Tower, Cherokee Six One Tango going to Ground.

Twr: *Cherokee Six One Tango, Roger. Contact Ground.*

This clears you to leave the tower frequency. Remember, stay on the tower frequency until a change is authorized.

OTHER TRAFFIC PATTERN COMMUNICATIONS

Because of spacing, the tower wants greater separation between you and the aircraft ahead of you:

Twr: *Cherokee Six One Tango, extend your downwind for spacing.*

You: Roger. Cherokee Six One Tango. Will you call my base? [Meaning, "Will you tell me when I can turn to the base leg?"]

Twr: *Cherokee Six One Tango, affirmative.*

Twr: *Cherokee Six One Tango, turn to base. You're number two to land behind the Aero Commander on final.*

You: Roger, and we have the Commander [or "negative contact on the Commander"], Cherokee Six One Tango.

You're on final, 100 feet above touchdown, and an unauthorized aircraft or ground vehicle ventures onto the runway. To avoid a confrontation, the tower issues a command:

Twr: *Cherokee Six One Tango, go around!*

You: [No response is necessary. Pour on the coals and initiate the go-around procedure. Don't argue; don't debate. Just do what you're told.]

You've been shooting touch-and-gos but would now like the option of making a touch-and-go, stop-and-go, or a full-stop landing. Make the request on the downwind leg so the tower can say yea or nay, based on the existing traffic:

You: Cedar Tower, Cherokee Six One Tango requests the option.

Twr: *Cherokee Six One Tango, cleared for the option.*

The option approach is especially useful during flight instruction to maintain an element of surprise for the student, because go-arounds and missed instrument approaches are also permitted as options.

On final approach, decide (or have your instructor decide) how you'll end the approach. No need to tell the tower. You're cleared for whatever you decide.

When you've had enough for the day, advise the tower of your intentions—again on the downwind leg:

You: Cedar Tower, Cherokee Six One Tango will be full stop this time.

Twr: *Cherokee Six One Tango cleared to land.*

You: Cleared to land, Cherokee Six One Tango.

At this point, the tower makes a request of you:

Twr: *Cherokee Six One Tango, can you land short and turn off on Runway Two One?*

You: Affirmative, Cherokee Six One Tango. [Assuming you can comply with the request.]

Don't just come back with a "Roger." Can you or can't you? The response is either "Affirmative" or "Negative."

A TOWER BUT NO GROUND CONTROL OR ATIS ON THE FIELD

As I indicated in chapter 6, most tower-controlled airports provide Ground Control service. At many of the smaller and less active fields, however, the tower controller is often responsible for both ground and flight movements. Since the sectional charts themselves do not indicate the presence or absence of the service, you have to refer to the table on the inside of the sectional cover flap. If you're at a tower-operated field that does not have Ground Control, merely contact the tower just as if there were a GC. The controller will then provide the necessary taxi instructions.

Occasionally, even when there is Ground Control, this can happen: You've just landed and are clear of the runway. Rather than telling you to contact Ground, the tower assumes the taxiing-control responsibility.

Twr: *Cherokee Six One Tango, taxi to the ramp.*

Or if there are several locations where you could park, it might be this:

Twr: *Cherokee Six One Tango, where do you want to park?*

You: Hangar Four, Tower.

Twr: *Roger, Six One Tango, taxi to Hangar Four.*

You: Roger, Six One Tango.

The assumption of the Ground Control role by the tower—if it does happen at all—occurs when the area traffic is light and the controller has the time to issue brief instructions. Just don't be surprised if the tower assumes the responsibility. The possibility of it, however, is another reason why you don't change frequencies without ATC approval.

As to the ATIS not being on the field: You know from chapter 5 that both its presence and frequency are indicated on the sectional and in the *A/FD*. In general, that tape-recorded service is available at most, but not all, tower-controlled airports.

Where you are more likely not to find ATIS is at those towers manned by nonfederal contractors. These are "civilians," if you will, who represent private firms that have contracted with city governments to provide the traffic-controlling service at the local airports. In the vast majority of such cases, and for whatever reason, ATIS is not part of that service. Frequently, however, these "NFCT" (Non-Federal Control Tower) fields are near major airports that do have ATIS that pilots can monitor for a fairly accurate picture of conditions at whatever non-ATIS field they're going to operate in.

Fine, but how do you know if it's a nonfederal tower? Again, the sectional tells you, both on the chart itself by the airport symbol and on the inside cover flap. On the chart, the indication is "NFCT" located in front of the tower frequency. If you'll check the Pendleton, Oregon, example in FIG. 9-1 again, you'll see what I mean. On the flap, a thin "NF" follows the airport name.

If you're at an NFCT field and there's no ATIS, the tower will give you the basic operational data—wind, direction, velocity, altimeter setting, and the active runway. That's all, barring unusual weather conditions in the airport vicinity. You now have "the numbers" essential for a takeoff or a landing. To get this data, simply request it in your initial call:

You: Billard Tower, Cherokee One Four Six One Tango is over Perry Dam at four thousand five hundred for landing. Request the numbers.

Twr: *Cherokee One Four Six One Tango, enter left base for Runway One Eight. Winds two zero zero at one five. Billard altimeter three zero one two. Report entering base leg.*

You: Roger, report entering left base for One Eight. Cherokee Six One Tango.

The procedure is exactly the same prior to taxiing for takeoff when there is no ATIS. Merely tell Ground (or the tower, if there is no Ground Control) where you are, that you're ready to taxi, followed by "Request the numbers." Either will give you the essential data.

OBTAINING SPECIAL VFR CLEARANCE

Back in the Flight Service chapter, I discussed obtaining Special VFR clearance when operating within a Control Zone under less than VFR conditions with no tower on the field. Now let's assume that the same conditions exist at an airport which has a tower. The procedures and calls are basically the same as those when a Flight Service Station is involved.

First, check the ATIS and then contact Ground Control and request the SVFR:

You: Municipal Ground, Cherokee One Four Six One Tango at Jet Air with Information Kilo. Request Special VFR southbound.

GC: *Cherokee One Four Six One Tango, taxi to Runway Three Three. Clearance on request.* [This means that Ground is requesting clearance for you from the Air Route Traffic Control Center. It does not mean that the clearance is available to you on your request.]

You: Roger. Taxi to Three Three, Cherokee One Four Six One Tango.

GC: [In a few minutes] *Cherokee One Four Six One Tango, clearance when ready to copy.*

You: [Assuming you are free to copy] Cherokee One Four Six One Tango ready.

GC: *ATC clears Cherokee One Four Six One Tango to exit the Municipal Control Zone to the south. Maintain Special VFR conditions at or below two thousand while in the Control Zone. Report leaving the Control Zone.*

You: Understand Cherokee One Four Six One Tango cleared Special VFR to depart south, maintain two thousand or below while in the Zone, and report clear of the Zone.

GC: *Cherokee Six One Tango, readback correct. Contact Tower.*

You: Roger, Cherokee Six One Tango.

When you've finished the pretakeoff check and are ready to go, switch to the tower, which will already be aware of your clearance. Thus the following:

You: Municipal Tower, Cherokee One Four Six One Tango ready for takeoff. Special VFR southbound.

Twr: *Cherokee Six One Tango, Roger. Cleared for takeoff. Report when clear of the Zone.*

You: Wilco, Cherokee Six One Tango.

Perhaps instead of departing, you want to land at Municipal. Through ATIS or any other source, you find that the field is below VFR limits but adequate for a Special VFR. Before entering the Control Zone, call the tower:

You: Municipal Tower, Cherokee One Four Six One Tango is over Deep Lake, level at three thousand, squawking one two zero zero with Information Xray. Request Special VFR for landing Municipal.

Twr: *Cherokee One Four Six One Tango, Roger. Remain clear of the Control Zone until further advised.*

Now circle, slow down, or do whatever is necessary to remain outside the Zone until you hear from the tower again:

Twr: *Cherokee Six One Tango, Municipal Tower. Clearance when ready to copy.*

You: Cherokee Six One Tango ready.

Twr: *Cherokee One Four Six One Tango is cleared to enter the Control Zone east of Municipal. Maintain Special VFR conditions at or below two thousand five hundred. Squawk one two zero five. Report right base for Runway Three Five.*

You: Roger. Cherokee One Four Six One Tango cleared to enter the Control Zone east of Municipal. Maintain Special VFR conditions at or below two thousand five hundred. One two zero five and report right base for Three Five.

Twr: *Cherokee Six One Tango, readback correct.*

You: Tower, Cherokee Six One Tango turning right base for Three Five.

Twr: *Cherokee Six One Tango, Roger. Cleared to land, Runway Three Five. Winds three two zero degrees at five.*

Of course, if the airport offers Approach Control, contact Approach for this clearance instead of the tower. You'll then be instructed when to contact the tower for landing clearance.

A FEW WORDS ABOUT DEVIATIONS

Something I've noticed, particularly with newer pilots, is the reluctance to ask the tower for traffic pattern deviations when another runway or approach is more practical. There is no emergency, no danger of violating a VFR regulation, but a different pattern would be more practical or time-conserving. While the reluctance is understandable, controllers occasionally authorize such deviations if the volume of traffic permits. To be more specific:

There are two runways on the field, one is 19/01, the other 21/03. You've already learned from monitoring the ATIS that One Nine is the active runway, but the wind is from 220 degrees at 15 knots, gusting to 25. That could present a fairly stiff cross-wind, so you decide to ask the controller in your initial contact if 21 can be used:

You: Downtown Tower, Cherokee One Four Six One Tango is over Lake Quivera at three thousand, squawking one two zero zero with Information Charlie for landing. Is Two One available?

Twr: *Affirmative, Cherokee One Four Six One Tango. Ident and enter left downwind for Two One.*

You: Roger, left for Two One. Cherokee Six One Tango.

Did you push the Ident button, as the controller asked?

Another possibility:

You're coming in from the west and, again from the ATIS, you learn that One Nine is the active runway with a left pattern. To enter the downwind for that pattern, though, means you'd have to fly over the field, go east a few miles, descend, and then join the other downwind traffic at a 45-degree angle. All of which is both time- and fuel-consuming. ATC might approve a direct entry to a right base leg almost dead-ahead of your present position. Thus the query:

You: Downtown Tower, Cherokee One Four Six One Tango is over Wyandotte Lake at two thousand five hundred, squawking one-two-zero-zero with Information Delta for landing. Request right base, if possible.

Twr: *Roger, Cherokee One Four Six One Tango, Ident, and report entering right base for One Niner.*

You: Roger, report right base for One Niner. Cherokee Six One Tango.

A few minutes later:

You: Tower, Cherokee Six One Tango on right base for One Niner.

Twr: *Cherokee Six One Tango, cleared to land.*

You: Six One Tango cleared to land.

If the tower can't grant your request, no problem. The controller will then come back with something like this:

Twr: *Unable right base, Cherokee Six One Tango. Ident and fly over the field at two thousand three hundred for a left downwind for One Niner.*

Fine, but at least you tried. Now merely acknowledge the instructions and do what the tower has said.

Regardless of the airspace, controllers can be flexible when conditions permit. You know where you are and what's best for you, so don't be afraid to request a reasonable deviation. Just be sure, though, that it's reasonable. No one in his right mind would ask to land on Runway 01 at a busy airport when the wind is out of the south and all the traffic is using 19. Let's hope not, anyway.

CONCLUSION

Remember, the communications examples cited in this book basically conform to the FAA-approved phraseology. It's appropriate to note, however, that you'll occasionally hear some verbal shorthand that might be acceptable but not necessarily correct by FAA standards.

For example, the fact that a pilot is "Squawking one two zero zero" is being heard less and less—if at all. Or it might be reduced to "Squawking Twelve." Pilots—not controllers—sometimes shorten altitudes from "Level at two thousand three hundred" to "Level at two point three." The fact that the pilot has "...Information Alpha" is often shortened to "...with Alpha." The word "Information," being somewhat self-evident, is dropped entirely. And there are others. I recommend, though, that you master the approved radio language first. Get that down before the tendency to slip into some of the typical but incorrect pilot jargon becomes too inviting.

To sum up, and as I've tried to illustrate in this chapter, there's nothing complicated or mysterious about operating in an ATA (Class D) airspace. From a radio communications point of view, it's simply a matter of adhering to a few basic principles, the majority of which can be condensed this way:

Monitor the ATIS 20 miles or so out from the airport.

Monitor the tower frequency ahead of time to learn what you can about the traffic pattern and the volume of activity in the area.

Plan what you're going to say and how you're going to say it.

Listen before you press the mike button to be sure the air is clear.

Listen to what the controller tells you.

Clarify instructions, if you're not sure.

Listen for your call sign even after you've been given landing pattern instructions. The tower might have additional instructions for you.

Acknowledge instructions briefly and tersely.

Inform the controller of what you're doing or going to do if a deviation is required.

Request deviations from instructions if there is a more effective or efficient non-emergency alternative.

Communicate with confidence.

In essence, you and everyone else in the airport vicinity make up a team. The controller calls the plays and the players conform. If a given play won't work, or if there's a better play for a given circumstance, fine, but tell the controller/coach first and get permission to deviate before embarking on some imaginative digression from the game plan.

When all parties function as a moment-in-time team, traffic flows as it should. And when it flows as it should, the folks up in the tower can be the nicest folks in the world. Help them help you.

10
TCAs, ARSAs, and Nonradar Airports

The Cerritos, California, midair collision in 1986 accelerated or instigated a flurry of FAA regulatory changes designed to prevent recurrences of such an incident. The changes since then, primarily affecting Terminal Control Areas (TCAs) and Airport Radar Service Areas (ARSAs), have involved their structure, adoption of stricter Mode C requirements, and more stringent instructional requirements for solo student-pilot operations in the TCAs.

Coming now, as you already know from chapter 8, is the airspace reclassification which replaces the "TCA" and "ARSA" designations with "Class B" and "Class C" airspaces. So changes have been, are, and probably will be, par for the course as the FAA continues to seek ways to increase safety in high-density traffic areas.

FIRST—A FEW WORDS ABOUT THE "DYING" TRSA

Speaking of change, one current terminal designation, the "Terminal Radar Service Area," or "TRSA," is not mentioned in the *Federal Register* reclassification publication, even though a few medium-sized airports around the country bear that label. TRSAs were quite prevalent in the pre-ARSA days, but they are gradually, perhaps rapidly, disappearing, either being upgraded to an ARSA (Class B) or down to an ATA (Class D) airspace. Those remaining, however, can be easily spotted on the sectional by large solid magenta-colored circles surrounding the primary airports.

Besides their structure, the major difference between a TRSA and an ARSA is that use of the TRSA Approach Control facility for radar service (or Stage III) is not required, nor is a Mode C transponder. Both, you'll remember, are essential when operating in a TCA or an ARSA.

Because of their phase-out, I won't spend time on the TRSA or its radio procedures. You might keep in mind, though, that if you go VFR into a TRSA airport and choose to request Stage III, the radio procedures are very similar to those in an ARSA as well as with an ATA tower. You still have to contact the tower outside the ATA in a TRSA, but otherwise the use of radar service is up to you.

The thrust of this chapter, then, is to review the structure and operating requirements in the TCA (Class B) and ARSA (Class C) environments. In the process, and just as I have already done, I'll inevitably refer to the Approach and Departure Control facility. Don't let these references confuse you, because I'll discuss that facility and its responsibilities in chapter 11, along with the correct radio phraseology. Let's say for now that when you contact Approach or Departure, you're not talking to an ATA tower controller but to another controller who has responsibility for the area out of and beyond the 5-sm airport radius.

Assume that you want to go into an airport that has a TCA, an ARSA, or one that has none of those but does provide radar Approach and Departure Control service. What is mandatory or optional in each situation? What do you say, and when do you say it?

If you're at all typical of the average noninstrument-rated pilot, you might be a bit confused as to your responsibility. (Perhaps *uncertain* is a better word.) Confusion or uncertainty keeps many pilots from venturing into controlled traffic areas. Instead, they opt for the smaller, less-convenient airports that pose no challenge to their communication expertise. TCAs and ARSAs are a bit too much—a bit too threatening.

It's the fear of the unknown, the feeling of insecurity, that causes pilots to detour miles around a TCA or put down in some remote airfield far from their ground destination. Insecurity is natural when the unknown confronts us, but insecurity can be conquered through education and practice. In this chapter, and the ones that follow, I'll give that hypothesis a try.

TERMINAL CONTROL AREA (TCA) (CLASS B AIRSPACE)

The traditional description of a TCA (Class B) is that of an upside-down wedding cake, each layer having a bottom (a floor) and all levels having a common ceiling. The core always surrounds the primary airport within the TCA, with the higher levels fanning out from the core.

Figure 10-1 is a generalized side view of a TCA. A top view is shown in FIG. 10-2. These are simplistic sketches, because TCAs are not always uniform in size, shape, or floor and ceiling altitudes. Nevertheless, the sketches reflect the fundamental concept.

At the time of this writing, there are 29 TCAs located at the following largest and busiest airports. Those with an asterisk (*) identify the nine busiest TCAs which have special pilot requirements.

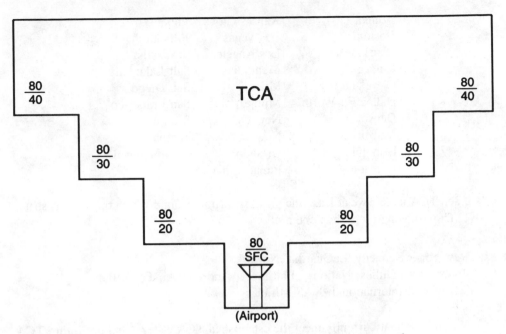

Fig. 10-1. *The TCA is typically described as an upside-down wedding cake.*

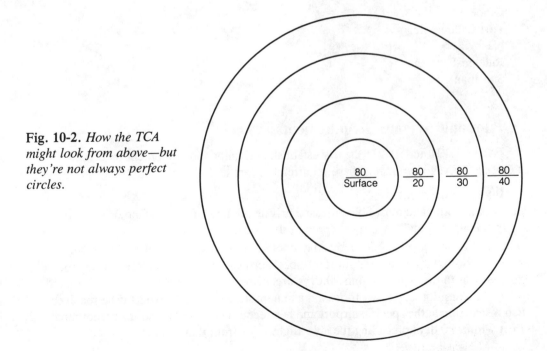

Fig. 10-2. *How the TCA might look from above—but they're not always perfect circles.*

Atlanta*	Kansas City	Phoenix
Boston*	Las Vegas	Pittsburgh
Charlotte	Los Angeles*	St. Louis
Chicago*	Memphis	Salt Lake City
Cleveland	Miami*	San Diego
Dallas/Ft. Worth*	Minneapolis	San Francisco*
Denver	New Orleans	Seattle
Detroit	New York*	Tampa
Honolulu	Orlando	Washington, DC*
Houston	Philadelphia	

Each TCA must have at least one primary airport around which the TCA is structured. Those having more than one are:

New York: Kennedy, LaGuardia, Newark
Washington: Dulles, National, Baltimore, Andrews Air Force Base
San Diego: International, NAS Miranar

Six airports that currently meet the established TCA criteria and are future TCA candidates:

Chicago Midway
Cincinnati
Fort Lauderdale
Nashville
Raleigh-Durham
San Juan

Pilot and Avionics Requirements

FARs 91.131 and 91.215 clearly establish both pilot and equipment requirements to enter a TCA. In summary, these requirements are as follows:

Pilot requirements

The pilot in command must hold at least a private pilot certificate to land or take off at any of the asterisked TCA airports previously listed.

A student pilot on a solo flight may operate in TCA airspace if he has received ground and flight instruction for that specific TCA and has received a logbook endorsement to that effect within 90 days preceding the solo flight.

A student pilot may take off or land at a nonasterisked TCA airport if he received such instruction at the specific airport and has received a logbook endorsement to that effect within 90 days preceding the solo flight to or from that airport.

Avionics requirements

- operable VOR or TACAN receiver.
- operable two-way radio with frequencies appropriate for the TCA.
- operable Mode 3/A 4096-code transponder (or Mode S transponder) with Mode C altitude reporting capability. This equipment is also required in all airspace within 30 nautical miles of a TCA primary airport (FIG. 10-3) and from the surface up to and including 10,000 feet msl. (See also FIG. 10-6.)

Participation: Voluntary or Mandatory?

When entering or departing a TCA, you have no choice. Participation is *mandatory*. FAR 91.131 makes this very clear:

No person may operate an aircraft within a terminal control area . . . unless that person has received appropriate authorization from ATC prior to operation of that aircraft in that area.

This should need no further elaboration. The dictum is ample warning to those who would intentionally or carelessly bust the TCA.

Identifying a TCA

TCAs are easy to identify on the sectional because their location is always outlined by a heavy blue square or rectangle that encompasses the entire area. Figure 10-3, the Kansas City TCA, illustrates the square as well as other features common to most TCAs. It should be noted again, though, that all TCAs are not the same. Their basic structures are similar, but their individual designs and vertical limits can vary considerably.

The actual TCA in this and all cases is not the square or rectangle but rather the solid blue circles or their irregular cutouts that surround the primary airport. The innermost circle represents the TCA's core, and, referring again to FIG. 10-3, just to the left of the runways you'll see "80/SFC." This means that the core rises from the airport surface to 8,000 feet msl, which is the ceiling of the entire TCA. Each succeeding outward circle defines the next layer or shelf, along with its ceiling and floor, as 80/24, 80/30, and so on. (These numbers should be read in thousands of feet: 8,000/2,400, etc.)

A point to note is that the floor of a given shelf is not always the same. For example, starting at the 12 o'clock position and going clockwise (FIG. 10-3), the outermost shelf first has a floor at 4,000 feet, then 5,000 feet, 4,000, 5,000, 4,000, and finally 5,000 feet. These varying floor levels are separated by the solid lines that distinguish one altitude from the next.

You should be alert to these variations for this reason: Suppose that you're going VFR from the Johnson County Industrial Airport (bottom center of FIG. 10-3) to the

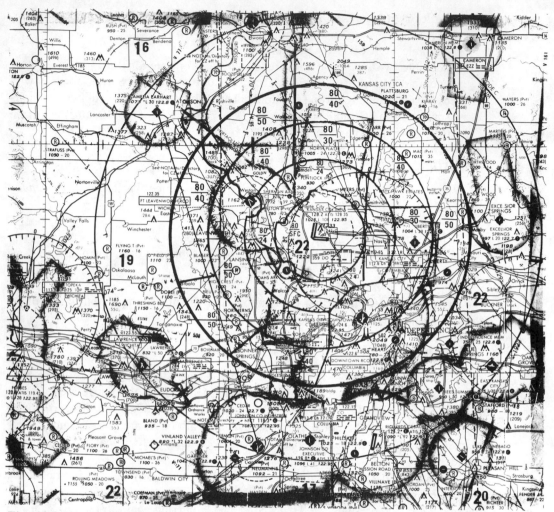

Fig. 10-3. *The large blue square makes TCA (Class B) identification on the sectional easy. The thicker circles represent the TCA itself, while the thin circle surrounding the entire TCA is the 30-nm-radius veil within which Mode C transponders are required.*

Atchinson Amelia Earhart Airport (upper left). A direct line, as plotted on the chart, will take you through the outer TCA shelf. But you don't want to enter the TCA, so you cruise at 4,500 feet to stay well below the shelf's 5,000-foot floor (and also to conform to the VFR altitude regulations above 3,000 feet). If you're not careful, though, and are still at 4,500 feet, you'll soon be in the 4,000-foot floor of that same shelf. Now you've done it! You've busted the TCA without radio contact or Approach Control approval.

Unless you're very lucky, Approach, noting the illegal penetration, will track your airplane by radar until you touch down at Atchinson. At that point, you can expect a telephone call from an Approach Control supervisor and the subsequent likelihood of some sort of disciplinary action. All because you didn't do enough preflight planning or disregarded what the sectional told you about the TCA.

The TCA Veil

Another feature that FIG. 10-3 illustrates is the limit of the Mode C *veil*. This veil is the thin blue line that circles the entire TCA and identifies the 30-nautical-mile radius from the primary airport.

Everywhere within this circle, Mode C transponders are required—except for those airports that are located no more than 2 nm inside the circle, which usually places them beyond the TCA's Approach Control radar coverage. Also, these must be airports from which aircraft can leave or enter the veil in a direct line and, in the process, not cause the pilot to violate FAR 91.119 (minimum safe altitudes).

In essence, FAR 91.119 states that no aircraft may operate below 1,000 feet agl over any congested area, city, town, settlement, or open-air assembly of persons. In other words, there are no obstructions that would require flight deviations or rule-violations and thus force the non-Mode C aircraft to operate within the veil, except for the traffic pattern, before the pilot could get out beyond the 30-mile limit. And exit that veil immediately he must, if he doesn't have Mode C.

One other aspect of these "fringe" airports is that the maximum operating altitude is limited. In the Kansas City veil, for example, non-Mode C aircraft must remain below 1,000 feet agl in the traffic pattern and when exiting or entering the airport area. At Cleveland, it's 1,300 feet agl; Memphis is 2,500 feet; Orlando, 1,400. The altitudes differ from TCA to TCA, but they are not indicated on the sectional. They can be found, however, in the *Airman's Information Annual* (*AIM*) Tables 3-30 (1) through 3-30 (25), under the controlling TCA.

While the focus so far has been on the sectional as the principal identification source of a TCA (Class B), don't overlook the *Airport/Facility Directory*. As FIG. 10-4 illustrates, the *A/FD* provides a wealth of information about the particular airport, plus, in this case, that it's a TCA. Before going into any airport—TCA (Class B) or otherwise—reference to an *A/FD* is almost essential, because only there will you find the crucial data that you'll need, both in the air and on the ground.

Also, before flying into or near a TCA, be sure to equip yourself with the current VFR Terminal Area Chart (TAC) for the specific TCA. The TAC provides an enlarged and more detailed depiction of the TCA area, prominent landmarks, reporting points, Approach Control frequencies (based on your position), and other alerting and identifying features. It helps not only in navigating through the TCA, but also in avoiding the TCA altogether.

Entering the TCA

As previously indicated, I'll give examples of the radio calls themselves in the next chapter. Right now, let's just establish the rules, with the assumption, of course, that you have met the FAR pilot qualifications for TCA operations summarized earlier.

```
KANSAS CITY INTL    (MCI)   15 NW    UTC-6(-5DT)   39°17'57"N 94°43'04"W                    KANSAS CITY
   1026    B    FUEL 100LL, JET A    LRA    ARFF Index C                                    H-2E, 4G, L-6H, A
   RWY 01L-19R: H10801X150 (ASPH-GRVD)    S-100, D-200, DT-350    HIRL CL    0.3% up S              IAP
      RWY 01L: MALSR. TDZ.           RWY 19R: ALSF2 TDZ. Rgt tfc.
   RWY 09-27: H9500X150 (ASPH-GRVD)    S-75, D-125, DT-180    HIRL CL
      RWY 09: MALSR. TDZ.            RWY 27: REIL. VASI(V4L)—GA 3.0° TCH 58'. Rgt tfc.
   AIRPORT REMARKS: Attended continuously. Waterfowl on and in vicinity of arpt Oct 1-Dec 15 and Apr 1-May 30. New
      parallel Rwy 01R-19L under construction 6500' E of Rwy 01L-19R. Will will not be open until Fall of 1992.
      When using high-speed exits C3, C5, C6, continue until first parallel taxiway, then use extreme caution when
      turning in excess of 90°. Prior approval required to park at airline gate areas. High-speed Taxiways A4, A6, A8,
      C2, C6 and C7 grooved within 10' of both edges. Maximum gross weight limiting for Taxiway B8 and postal apron
      is S-45,000 and D-60,000. Rwy 09-27 B747 max gross weight of 460000 lbs, L1011 260000 lbs, DC-10-30
      315000 lbs. Landing Fee. Flight Notification Service (ADCUS) available.
   WEATHER DATA SOURCES: LLWAS.
   COMMUNICATIONS: ATIS 128.35 (816-243-3853)    UNICOM 122.95
      COLUMBIA FSS (COU) TF 1-800-WX-BRIEF. NOTAM FILE MCI
      RCO 122.65 122.1R 112.6T (COLUMBIA FSS)
   ®APP CON 132.95 (010°-190°) 120.95 (191°-009°) DEP CON 126.6 (010°-190°) 124.7 (191°-009°)
      INTERNATIONAL TOWER 128.2 125.75    GND CON 121.8 121.65    CLNC DEL 135.7
      TCA: See VFR Terminal Area chart.
   RADIO AIDS TO NAVIGATION: NOTAM FILE MKC.
      (H) VORTAC 112.6    MKC    Chan 73    39°16'46"N 94°35'28"W    273° 6.0 NM to fld. 1060/8E. HIWAS.
      DOTTE NDB (MHW/LOM) 359    DO    39°13'15"N 94°44'59"W    012° 4.9 NM to fld. NOTAM FILE MCI.
      HUGGY NDB (LOM) 242    RN    39°18'07"N 94°51'03"W    086° 6.2 NM to fld. NOTAM FILE MCI.
      ILS/DME 109.7 I-RNI Chan 34 Rwy 09 LOM HUGGY NDB
      ILS 110.5 I-DOT Rwy 01L LOM DOTTE NDB.
      ILS 109.1 I-PAJ Rwy 19R
```

Fig. 10-4. *How the* A/FD *identifies a TCA and the Clearance Delivery frequency.*

Using the Terminal Area Chart as the primary reference, always contact Approach Control on the frequency printed on the TAC well before you near the TCA shelf that you intend to enter. This means that the call should be made when you're about 15 miles out from that shelf, as illustrated in FIG. 10-5. Otherwise, on a busy day with a steady flow of radio communications, you might be about to enter the TCA before Approach can respond to your call. And entry without a clearance is a distinct no-no.

Once Approach has cleared you, you'll be given headings and altitudes to fly until Approach authorizes you to change to the tower frequency. You're then in, or almost in, the ATA, and the tower will give you the final landing instructions.

Let's set up another situation: The primary airport has sufficient traffic to warrant a TCA, but you want to land at another field that lies under the TCA but outside of the primary airport's immediate vicinity. St. Louis is a good example. The core of the TCA from the surface to 8,000 feet surrounds Lambert Field. About 20 miles west-southwest is Spirit of St. Louis, a tower-controlled airport. From a side view, looking northward, the TCA appears as depicted in FIG. 10-6. You're approaching Spirit from the west. Are you required to contact Approach? Yes, if your position and altitude would put you into any one of the floor/ceiling levels. No, if you stay below the floors. You must, however, contact Spirit's control tower before you enter its ATA.

So much for entering and landing in a TCA. How about taking off? Here are three situations.

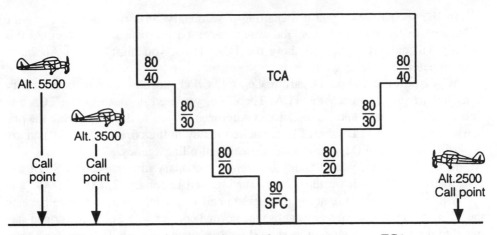

Fig. 10-5. *When to contact Approach Control if you want to enter a TCA.*

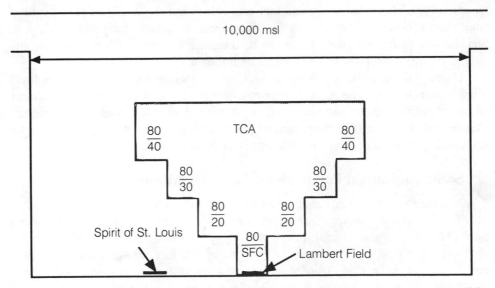

Fig. 10-6. *Spirit of St. Louis is an example of an airport that lies under, but not in, a TCA.*

In the first case, you're departing the primary TCA airport. You must receive a clearance from Clearance Delivery and, according to the instructions given you, you must contact Departure Control after takeoff.

All airports in TCAs and ARSAs have Clearance Delivery. Operating on its own frequency, its purpose is to communicate clearances in order to reduce congestion on the Ground Control frequency. Its availability at a given airport is identified in the "Communications" section of *Airport/Facility Directory* (FIG. 10-4).

In the second case, you're departing a secondary airport under—not in—the TCA, such as Spirit of St. Louis. You want to go east and cruise at 5,500 feet. As this altitude will quickly place you above the TCA floors, you must contact Approach Control before penetrating any floor.

Why Approach and not Departure Control? Good question. Leaving a secondary field you are going to *enter* the TCA. Therefore, you are *approaching* the TCA and thus under the control and surveillance of Approach. If you're departing from the primary airport, such as Lambert Field, you're already in the core of the TCA and are departing from it, so Departure is the logical controlling agency.

In the third case, you're taking off from a secondary airport such as Spirit and plan to fly locally or leave the area by staying well under the 2,000-, 3,000-, and 4,000-foot floors. There is no requirement to call Approach because you're beneath the TCA, not in it. You merely establish the normal contact with Spirit Tower and stay tuned to the tower frequency at least until you're outside the 5-mile ATA.

One more situation: You're on a cross-country and your route takes you directly through a TCA. You don't want to land at any airport within the area, however, and to go around the TCA would only add time and cost to the flight. If you're not sure of your radio procedures, you'll probably circumvent the whole thing and watch your fuel dwindle as the bill rises. TCAs can be transited, but only with permission, and only if you meet the pilot and equipment requirements.

If you're headed east, say from Kansas City to Cincinnati, at 7,500 feet, the direct route is over St. Louis. Being cost-conscious, you call St. Louis Approach 30 miles or so out, identify yourself, give your position, altitude, destination, squawk, and request clearance through the TCA. Approach takes it from there and notifies you when radar service is terminated once you're clear of the TCA on the east side.

Some Additional Points about TCA Operations

A couple of other points about operating in or near a TCA might be helpful.

The first is a rule change that becomes effective September 16, 1993, in association with the airspace reclassification. Currently, and until that date, VFR aircraft operating in a TCA must maintain a separation of 500 feet below, 1,000 feet above, and 2,000 feet horizontally from any cloud. If a controller's instruction would cause you to come closer to a cloud than those minimums, and thus violate that regulation, it is your responsibility to so advise him and take the necessary evasive action.

Under reclassification, that rule will change in the TCA (Class B airspace) and require VFR aircraft only to remain *clear of clouds*. The basic reason for the change is to minimize the disruption of the flow of other traffic in the TCA when the VFR pilot has to alter assigned altitude or course headings to comply with the current regulation. Even with the new ruling, though, VFR traffic must still avoid entering any cloud or cloud layer and must advise the controller before deviating from the assigned course or altitude.

A second point: If you're flying outside a TCA but under one of the floors, be sure that you have the current TCA airport altimeter setting and then stay at least 200 feet below the floor in smooth air. If the air is rough, 400 to 500 feet below is better. You don't want to find yourself suddenly bounced up or caught in an updraft that would thrust you into the shelf and thus the TCA—which could easily happen if you were skimming along only a few feet beneath the floor.

Third: Just as you can't enter a TCA without prior approval, IFR aircraft can't operate below one of the shelves, whether arriving or departing. You thus should have no concern about seeing a 747 pop up in front of you, even in a busy airline hub environment. What's good for the goose . . . .

Fourth: You want to go through a TCA but not land, so you contact Approach for entrance clearance and radar vectoring. In response, the controller denies the request—which is his right whenever a VFR aircraft is involved. It's the controller's responsibility to sequence and separate IFR aircraft, but if the weather is questionable or the volume of traffic heavy, his only choice might be rejection of your request. He just doesn't have time to handle a VFR operation under the circumstances.

Should denial be his response, don't argue, don't complain. He has his reasons and is under no obligation to explain those reasons. His word is final, and your alternative is cut and dried: Stay out and go around the whole TCA, even if it costs you time and fuel.

Further to that point, the weather might be fine and all the rest, but the controller can still deny TCA entrance if the pilot comes across as uncertain, confused, or unknowing. These folks on the ground are busy enough without having to deal with incompetence, and how you sound over the air could lead them to that conclusion about you—false though it may be.

To reject a VFR pilot's request for the last reason is perhaps rare, but it does happen. Conversely, most controllers do everything they can to help all pilots—VFR or otherwise. After all, many are pilots themselves, and they have no trouble putting themselves in the left seat.

AIRPORT RADAR SERVICE AREA (ARSA) (CLASS C AIRSPACE)

Between the regulatory controls of the TCA (Class B) and the ATA (Class D) airspaces comes the Airport Radar Service Area—the ARSA (Class C). Here, there are controls, to be sure, but getting into or out of an ARSA is easier, the structure is smaller, and the traffic volume less than in a TCA.

To qualify for ARSA consideration, the FAA has stipulated that the primary airport must have a Control Tower, be served by a radar Approach/Departure Control facility, and meet at least one of the following conditions:

- A minimum of 250,000 passenger enplanements a year, *or*
- at least 75,000 instrument operations a year, *or*
- at least 100,000 instrument operations a year at the primary and secondary airports included in the ARSA.

Identifying the ARSA and Its Structure

An ARSA (Class C airspace) is easily identified on the sectional by two thick slashed magenta circles that surround the primary airport (FIG. 10-7). For the most part, these circles are concentric, but, like the TCAs, there might be jogs or irregularities based on terrain or the proximity of other non-ARSA airports. The *Airport/Facility Directory* also identifies ARSAs, as illustrated in FIG. 10-8.

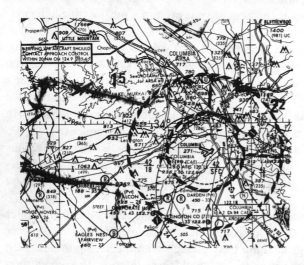

Fig. 10-7. *The slashed concentric circles (magenta on the sectional) identify an ARSA.*

COLUMBIA
 COLUMBIA METROPOLITAN   (CAE)   5 SW   UTC−5(−4DT)   33°56′25″N 81°07′10″W   ATLANTA
   236   B   S4   FUEL 100LL, JET A   OX 1, 3   TPA—See Remarks   AOE   ARFF Index C   H-4I, 6F, L-20F, 27B
   RWY 11-29: H8602X150 (ASPH-GRVD)   S-72, D-200, DT-328, DDT-700   HIRL CL   IAP
   RWY 11: ALSF2. TDZ. Tree.   **RWY 29:** MALSR.
   RWY 05-23: H7000X150 (ASPH-GRVD)   S-72, D-200, DT-315   HIRL
   RWY 23: VASI(V4L)—GA 3.0° TCH 54′. Tree.
   AIRPORT REMARKS: Attended continuously. PPR arpt manager before acft transporting explosives lands at fld. Obtain
      permission during business hours 1330–2200Z‡ Mon–Fri except holidays 803–822-5000. All 180°turns on
      grooved surfaces prohibited. Fee for commercial aircraft over 15,000 pounds. Bird activity on and in vicinity of
      arpt. Opr of ultralight vehicles prohibited. TPA for propellar acft 1236(1000); TPA for turbprop 2036(1800).
      NOTE: See SPECIAL NOTICE—Simultaneous Operations on Intersecting Runways.
   COMMUNICATIONS: ATIS 120.15   UNICOM 122.95
      ANDERSON FSS (AND) TF 1–800–WX–BRIEF. NOTAM FILE CAE.
      RCO 122.1R 114.7T (ANDERSON FSS)
   ®APP/DEP CON 118.2 (110°-289°) 124.9 (290°-109°)
      TOWER 119.5   GND CON 121.9   CLNC DEL 119.75
      ARSA ctc APP CON
   RADIO AIDS TO NAVIGATION: NOTAM FILE CAE.
      (H) VORTAC 114.7   CAE   Chan 94   33°51′26″N 81°03′15″W   329° 6.0 NM to fld. 410/02W.
      MURRY NDB (LOM) 362   CA   33°58′01″N 81°14′42″W   109° 6.5 NM to fld.
      ILS 110.3 I-CAE Rwy 11. LOM MURRY NDB.
      ILS 108.3 I-VYK Rwy 29.
      ASR

Fig. 10-8. *How the* A/FD *identifies an ARSA.*

Another helpful feature in recent sectionals is inclusion of a wide blue square, similar to TCAs, that encompasses 30 to 40 nms of the ARSA area. This area is enlarged on the inside cover of the chart to provide more geographic detail than otherwise possible—much like the TCA Terminal Area Chart. This doesn't apply to all ARSAs, however. For example, the Cincinnati sectional boxes and enlarges the Cincinnati and Dayton ARSAs, but not Columbus. And the same is true of the many other ARSAs: No box, no enlargement. The feature is not yet universal, but that doesn't detract from the easy identification of the airspace.

Structurally, the ARSA is simple. It has only two rings, or circles, with the inner circle rising approximately 4,000 feet above the primary airport. The radius of the inner circle extends 5 nautical miles from the airport. The second, or outer, circle has a 10-nautical-mile radius, with varying floors and the same approximate 4,000-foot agl ceiling.

Note in FIG. 10-7 that the 5-nm radius ARSA inner circle is slightly larger than the 5-sm radius Control Zone and the 5-sm radius Airport Traffic Area, uncharted.

So far, the ARSA's shape is rather like a TCA, except that the ARSA is considerably smaller in radius. One design feature, however, makes ARSA different; it's called the *outer area*. This area begins at the edge of the outer circle and extends out for another 10 miles, giving the ARSA, in effect, a 20-nm dimension. In addition, the outer area rises from the lower limits of radio and radar coverage to altitudes of 10,000 to 12,000 feet agl, or, as the AIM explains it," . . . up to the ceiling of the Approach Control's delegated airspace." Figure 10-9 illustrates the typical horizontal configuration of an ARSA. Referring back to FIG. 10-7, though, you'll notice that the outer area is never depicted on a sectional.

Pilot and Equipment Requirements

ARSAs have no special pilot requirements. The only equipment needed is a two-way radio capable of sending and receiving on the ARSA frequencies, and a Mode C transponder for operations within an ARSA and above it, up to and including 10,000 feet msl.

Pilot and Controller Responsibilities in the ARSA

From the pilot's point of view, and according to FAA regulations, no "clearance" into the ARSA is required. As FAR Part 91.88 states:

Arrivals and Overflights. No person may operate an aircraft in an airport radar service area unless two-way radio communication is established with ATC prior to entering that area and is thereafter maintained with ATC while within that area.

What this means is that if you call an ARSA Approach Control and all that the controller says is, "Cherokee One Four Six One Tango, stand by," you have, by that acknowledgment, established "two-way radio communications" and can thus enter

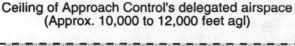

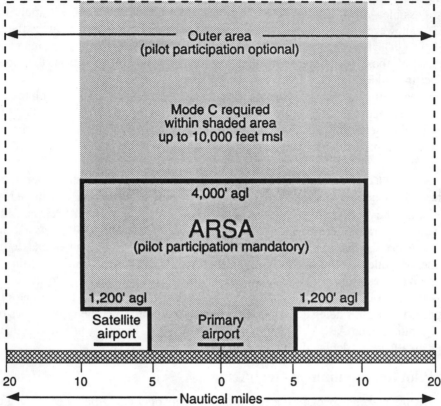

Fig. 10-9. *The horizontal and vertical dimensions of the typical ARSA.*

the ARSA. On the other hand, if the controller comes back with, "Aircraft calling Columbia Approach, stand by," you are *not* cleared into the ARSA. Your aircraft call sign must be included in the call to authorize entrance—even if the controller tells you to "stand by."

The same radio communication requirement applies to departing aircraft, but how that's done when leaving the primary ARSA airport I'll cover later. Also, aircraft departing a satellite airport within the inner circle must establish radio contact with the ARSA Departure Control as soon as possible after takeoff and maintain that contact while in the ARSA.

For those familiar with ARSA operations, the airspace reclassification changes nothing. Dimensions, altitudes, and regulations are the same as before. The matter of cloud separation, however, is worth noting. While the VFR pilot only has to remain "clear of clouds" in a TCA, he must abide by the "500 below, 1,000 above, and

2,000 horizontal" separation in an ARSA—which is the current regulation, and will continue to be effective after September 16, 1993.

Once radio contact has been established, ATC will provide the following services within the 10-mile ARSA radius:

- Sequencing of all aircraft arriving at the primary airport.

- Maintaining standard separation of 1,000 feet vertically and 3 miles horizontally between IFR aircraft.

- Traffic advisories and conflict resolution between IFR and known VFR aircraft so that the radar targets do not touch, or 500 feet vertical separation. (ATC will issue advisories or safety alerts if the targets are appearing to merge and the "green between" the targets on the scope is diminishing.)

- Issuing advisories and, if necessary, safety alerts to VFR aircraft.

The last point warrants clarification for those operating VFR. Unlike the TCA, ATC is not required to *separate* VFR aircraft in the ARSA, but may do so in certain locations if the density of traffic indicates the need. Separation involves whatever vectoring or altitude changes are necessary to maintain a given distance between aircraft. *Advisories* alert the pilot to bearings, approximate distance, and altitude in relation to other aircraft that might pose a potential safety problem. It is then the pilot's responsibility to scan the skies, locate the traffic, and advise ATC when the traffic has been spotted.

A *safety alert* is issued when conflict between two aircraft, or an aircraft and a ground obstruction, seems imminent. When transmitted, it means that the receiving pilot should change course, altitude, or possibly both, now. The situation is reaching emergency proportions. Alerts will be few, however, if advisories are heeded and the pilot is attentive to what's going on outside the cockpit.

The Outer Area

The outer area is not regulated airspace—not really part of the ARSA—therefore two-way radio communication in the outer area is not mandatory. Although controllers are learning to live with it, this outer area has been a source of some dispute. It can be at once a plus and a minus for pilots, while at the same time placing an added workload on the controller. To explain:

Let's say that you leave the primary airport and go to the outer area to practice maneuvers. You've maintained radio contact with Approach (actually Departure Control) and while you're in the outer area, the controller must provide the advisories and alert services just as though you were in the inner or outer circles. But, as soon as you hit the outer area, you have the right to request that those services be terminated. Unless you do, the controller has no alternative but to continue them.

Another example: You leave a satellite airport in the outer area and intend to stay in that area and not enter the ARSA. If you want the advisory and alert services, you must specifically request them. Otherwise, you won't hear from Approach at all.

Said simply, the VFR pilot is in the driver's seat. He can request, not request, or terminate the services in the outer area at will. The controller has no such prerogative.

The potential—or real—problem lies in the fact that a whole lot of aircraft could be milling around and requesting radar service in the outer area, which extends up to about 10,000 feet. ATC, having to provide that service to these aircraft, could be so busy that it would be forced to temporarily deny landing or transiting aircraft entry into the ARSA.

The policy states that clearance into the ARSA is not required, but some controllers have found that they have had to tell pilots who are at the fringes of the outer circle (not the outer area) to "remain clear." The volume of activity in the outer area has, at times, created such an additional workload that service to those wanting to enter the ARSA has been delayed. Consequently, the policy can benefit one group and adversely affect another—with the controller caught in the middle.

As *AIM* says, "While pilot participation in [the outer] area is strongly encouraged, it is not a VFR requirement." Not debating the value of radar service or your right to receive it, it would be considerate of others to terminate the service if and when you're going to fly in and around the outer area. If there's a valid reason for wanting the service, fine, but too many pilots who require advisories only consume airtime, add work for the controller, and possibly delay others' entry into the ARSA.

So, let's summarize the ARSA concept this way:

- The only equipment required is a two-way radio capable of communicating with ATC and a Mode C transponder for operations within the ARSA and above it, up to and including 10,000 feet msl.

- Radio contact with ATC must be established before entering the ARSA.

- ATC assumes that ARSA service is wanted as soon as radio contact is established. If you call Approach and the controller responds to you with "Cherokee One Four Six One Tango, stand by," you have satisfied the ARSA communications requirements and may enter the ARSA, unless you are told to remain clear.

- Participating aircraft must comply with all ATC vectors, instructions, and other directions.

- ATC will provide standard separation between IFR aircraft (1,000 feet vertically, 3 miles horizontally); minimum VFR-IFR separation (maintaining "green between" on the radar screen or 500 feet vertically); and advisories and, if necessary, safety alerts to VFR aircraft. No separation is normally provided between VFR aircraft.

- ATC will sequence landing aircraft in daisy-chain fashion and turn each aircraft over to the local Control Tower (primary or secondary airport) for final landing instructions.

- Outer area services are provided as long as radio communications are maintained. They are terminated only when the pilot advises that ARSA services are not desired.

- ARSA service to aircraft landing at satellite airports within the ARSA is discontinued when the pilot is instructed to contact that airport's Control Tower.

- Aircraft departing a satellite airport will not receive ARSA service until they have established radio communications and are radar identified.

- Participation is mandatory for any aircraft landing, departing, or transiting the 10-mile ARSA radius (except under the floor of the outer circle). It is voluntary if landing, departing, or operating within the outer area.

Conclusion

That's the story of ARSAs—the sort of junior TCAs in the airspace system. There are currently 121 such Class C airspaces around the country, including Air Force bases, so you're more than likely to encounter one if you do any flying beyond the Class D ATA or an uncontrolled airport. It's a good idea to know what they are, their structure, and their requirements.

One problem, however, accompanies operating in an environment such as an ARSA or a TCA. It's the problem of complacency. You're under radar scrutiny and a controller is giving you headings and altitudes to fly, as well as traffic advisories. With that extent of ground guidance, the tendency is strong to keep one's head in the cockpit rather than scanning the skies to see what's going on outside.

As *AIM* puts it, in a rather lengthy sentence, in reference to ARSAs:

> This program is not to be interpreted as relieving pilots of their responsibilities to see and avoid other traffic operating in basic VFR weather conditions, to adjust their operations and flight path as necessary to preclude serious wake encounters, to maintain appropriate terrain and obstruction clearance, or to remain in weather conditions equal to or better than the minimums required by FAR 91.155.

AIM concludes, as I've stated before, that if a given ATC-directed heading or altitude is likely to compromise the pilot's responsibility with respect to other traffic, terrain, and so on, he should advise the appropriate ATC and obtain alternate instructions.

Said simply, don't be lulled into complacency just because a controller at a radar screen is telling you what to do. Keep your ears tuned to his instructions, but, above all, keep your eyes open for any condition that might become an unwelcome interruption of your flight. That's your job as a VFR pilot in these areas of controlled and sometimes voluminous traffic.

NON-TCA/ARSA RADAR APPROACH/DEPARTURE CONTROL AIRPORTS

Some ATA Class D airports that are not part of a TCA or an ARSA are equipped with their own Approach/Departure Control facilities. This fact, however, can't be

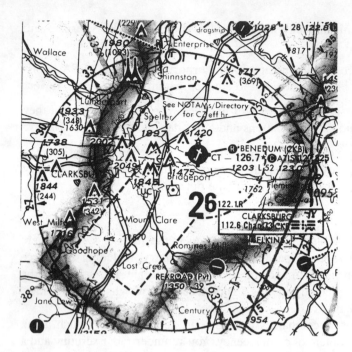

Fig. 10-10. *Clarksburg has no TCA, no ARSA, but it does offer radar Approach/Departure Control. The sectional, however, doesn't tell you that.*

determined from the sectional. As one example, take Clarksburg, West Virginia (FIG. 10-10).

From all appearances, Clarksburg is just a standard tower-only Class D airport. If you wanted to land there, however, a glance at the *Airport/Facility Directory* would tell you that radar Approach/Departure Control is located on the field. The clue is the symbol® , as shown in FIG. 10-11. The *A/FD* also indicates the availability of "Stage II" service, should you want it. Stage II is part of the radar service that includes traffic advisories and sequencing of VFR aircraft, but not their separation. The service is not mandatory, however. If you don't want it, all you have to do is tell the Approach Controller "Negative Stage Two" in the initial radio contact.

The *A/FD* excerpt in FIG. 10-11 also indicates that the Cleveland Air Route Traffic Control Center provides radar approach service to Clarksburg. It does this through its remote radar outlet, assuming the responsibility when the local Approach/Control facility is closed.

Airports that do not have their own Approach/Departure Control facilities—and there are many—can often rely on the services of a neighboring radar airport. As FIG. 10-12 illustrates, Baltimore provides the service for Gaithersburg, Maryland, and the Philadelphia Center does the same for Wilmington, Delaware (FIG. 10-13). So almost all of the Class D airspace airports are covered, as are many of those that are uncontrolled in the Class E airspace.

CLARKSBURG
BENEDUM (CKB) 0 NE UTC–5(–4DT) 39°17'44"N 80°13'44"W   **CINCINNATI**
  1203   B   S4   **FUEL** 80 100LL, JET A   OX 1, 2   ARFF Index A.   **H-4I, 6H, L-22F, 23D, 24E**
  RWY 03-21: H5198X150 (ASPH-GRVD)   S-70, D-90   HIRL   0.4% up NE   **IAP**
    RWY 03: REIL. VASI(V4L)—GA 3.44°TCH 59'. Trees.   **RWY 21:** MALSR. Trees.
  AIRPORT REMARKS: Attended continuously. PPR 24 hours for unscheduled air carrier operations with more than 30 passenger seats call arpt manager 304–842–3400. During hours twr clsd ACTIVATE MALSR—CTAF. Deer on and in vicinity of arpt. Ldg fee for all acft over 6500 lbs. Control Zone effective 1200-0400Z‡.
  WEATHER DATA SOURCES : LAWRS.
  COMMUNICATIONS: CTAF 126.7   **ATIS** 127.825(1200–0400Z‡)   **UNICOM** 123.0
    MORGANTOWN FSS (MGW) TF 1-800-WX-BRIEF. NOTAM FILE CKB.
    CLARKSBURG RCO 122.1R 112.6T (ELKINS FSS)
  Ⓡ**CLARKSBURG APP/DEP CON** 119.6 (West) 121.15 (East) (1200-0400Z‡)
  Ⓡ**CLEVELAND CENTER APP/DEP CON** 126.95 (0400-1200Z‡)
    CLARKSBURG TOWER 126.7 (1200-0400Z‡)   **GND CON** 121.9
    STAGE II SVC ctc **APP CON**
  RADIO AIDS TO NAVIGATION: NOTAM FILE CKB.
    CLARKSBURG (L) VOR/DME 112.6 CKB   Chan 73   39°15'11"N 80°16'05"W 042° 2.6 NM to fld. 1430/04W.
    ILS 109.3 I-CKB Rwy 21 Glide Slope unusable below 1600'. ILS unmonitored when twr clsd.
    ASR

Fig. 10-11. *The* A/FD *shows that Clarksburg has Approach/Departure Control, that the Cleveland Center provides the service when the Clarksburg facility is closed, and that Stage II Service is available.*

GAITHERSBURG
MONTGOMERY CO AIRPARK (GAI) 3 NE UTC–5(–4DT) 39°10'06"N 77°09'59"W   **WASHINGTON**
  539   B   S1   **FUEL** 100LL, JET A   OX 4   TPA—See Remarks   **L-24G, 28E, A**
  RWY 14-32: H4196X73 (ASPH)   MIRL   1.1% up SE   **IAP**
    RWY 14: REIL. VASI(V4L). Thld dsplcd 180'. Trees.
    RWY 32: REIL. APAP(PNIR)—GA 4.0° TCH 46'. Building. Rgt tfc.
  AIRPORT REMARKS: Attended 1300–0100Z‡. Unattended Thanksgiving, Christmas and New Years. ACTIVATE MIRL Rwy 14–32 and REIL Rwy 32—122.85. REIL Rwy 14 out of svc indefinitely. Noise abatement depart Rwy 32 turn rgt to at least 340°, refrain from Rwy 32 tkf between 0400–1200Z‡. TPA—1339(800), helicopter 1139(600).
  WEATHER DATA SOURCES: AWOS-3 128.275 (301) 977–2971.
  COMMUNICATIONS: CTAF/UNICOM 122.7
    LEESBURG FSS (DCA) TF 1–800–WX–BRIEF. NOTAM FILE DCA.
  Ⓡ**BALTIMORE APP/DEP CON** 128.7   **CLNC DEL** 121.6
  RADIO AIDS TO NAVIGATION: NOTAM FILE IAD.
    ARMEL (L) VORTAC 113.5   AML   Chan 82   38°56'04"N 77°28'01"W   053° 19.9 NM to fld. 297/08W.
    FREDERICK (T) VOR 109.0   FDK   39°24'44"N 77°22'32"W   154° 17.6 NM to fld. NOTAM FILE DCA.
    GAITHERSBURG NDB (MHW) 385   GAI   39°10'04"N 77°09'50"W   at fld. (VFR only) NOTAM FILE DCA.

Fig. 10-12. *Gaithersburg is an uncontrolled airport, but Baltimore provides Approach/Departure Control.*

"DESIGNATED" AIRPORT AREAS

So far, only two airports—Logan International in Billings, Montana, and Hector International in Fargo, North Dakota—fall under the "Designated" classification. For local reasons, the FAA has stipulated that Mode C transponders are required from the

WILMINGTON

NEW CASTLE CO   (ILG)   4 S   UTC-5(-4DT)   WASHINGTON
    39°40'43"N 75°36'25"W   H-6I, L-24G, 28F
  80   B   S4   FUEL 100LL, JET A,   OX 1, 2, 3, 4   LRA   ARFF Index A   IAP
RWY 09-27: H7165X150 (ASPH-GRVD)   S-90, D-140, DT-250   HIRL
  RWY 09: ODALS. Tree.   RWY 27: VASI(V4L)—GA 3.0°TCH 51'. Pole.
RWY 01-19: H7002X200 (ASPH-GRVD)   S-90, D-140, DT-250   HIRL
  RWY 01: ASLF1. Road.   RWY 19: VASI(V4L)—GA 3.0°TCH 58'. Tree.
RWY 14-32: H4594X150 (ASPH)   S-50, D-60   MIRL
  RWY 14: Tree.   RWY 32: VASI(V4L)—GA 3.0°TCH 27'. Sign.
AIRPORT REMARKS: Attended continuously. Birds on and in vicinity of arpt. When twr clsd ACTIVATE ALSF1 Rwy 01
  and ODALS Rwy 09—CTAF. Ldg fee for all acft except federal government and military. Rwy 09-27 no touch and
  go ldg for turbo jet 0400-1200Z‡. CLOSED to unscheduled air carrier operations with more than 30 passenger
  seats except one hour PPR, call 302-323-2680. Rwy 14-32 CLOSED to FAR Part 121 operators except 1 hour
  PPR call 302-323-2680. Extensive pilot training for fixed wing and helicopters. When terminal building clsd
  0400-1100Z‡ contact arpt security on 121.7 or 302-984-7520 (digital pager). Taxi lane section between Twy
  O and S limited to acft with wingspan of 49 ft or less. NOTE: See SPECIAL NOTICE—Simultaneous Operations on
  Intersecting Runways.
COMMUNICATIONS: CTAF 126.0   ATIS 123.95 (1200-0400Z‡)   UNICOM 122.95
  MILLVILLE FSS (MIV) TF 1-800-WX-BRIEF. NOTAM FILE ILG.
  DUPONT RCO 122.1R 114.0T (MILLVILLE FSS)
®PHILADELPHIA APP/DEP CON 118.35
  WILMINGTON TOWER 126.0 (1200-0400Z‡)   GND CON 121.7
RADIO AIDS TO NAVIGATION: NOTAM FILE ILG.
  DUPONT (L) VORTAC 114.0   DQO   Chan 87   39°40'41"N 75°36'27"W   at fld. 71/10W:
  HADIN NDB (LOM) 248   IL   39°34'52"N 75°36'52"W   013° 5.9 NM to fld.
  ILS 110.3 I-ILG Rwy 01 LOM HADIN NDB. ILS unmonitored when twr clsd.

Fig. **10-13.** *Although Wilmington, Delaware, has a Tower, Approach/Departure Control is Philadelphia's responsibility.*

surface to 10,000 feet msl within a 10-nautical-mile radius of these airports, excluding the airspace below 1,200 feet agl outside of their respective Airport Traffic Areas.

Besides factors of a particularly local nature, these airports, as well as any that may be so designated in the future, have high passenger traffic (200,000 enplanements annually), existing radar service, and substantial instrument operations. They are not yet, however, planned for an ARSA (Class C) classification.

CONCLUSION

As some of the foregoing information might be a little confusing, TABLE 10-1 summarizes the operating requirements of TCAs, ARSAs, and the other classes of airspace (even TRSAs) in the terminal environment. In only two cases is VFR pilot contact with Approach or Departure Control mandatory: When entering, departing, or transiting a TCA (Class B) or an ARSA (Class C). Going over, around, or under either requires no radio contact (or, as the term goes, "participation"). Of course, it's another matter if an airport has an operating Control Tower, and thus an ATA. Then approval to enter, leave, or transit the ATA is always required.

Although I've mentioned this before, the subject is so essential that it has to be stressed again: Operating in a radar-controlled environment in no way relieves the VFR pilot of the responsibility to maintain constant surveillance of the skies around

Table 10-1. Summary of airspace requirements

	Pilot participation required	Two-way radio communications required	Mode C transponder required[2]	VOR required	Special pilot requirements	Specific verbal "clearance" required	Sectional chart indication
TCA (Class B airspace)	✓	✓	✓[1]	✓	✓[3]	✓	Solid[4] blue lines
ARSA[5] (Class C airspace)	✓	✓	✓				Slashed magenta lines
ARSA Outer area	6	6	6				Not shown
TRSA	7	7					Solid magenta lines
Other radar Approach/Departure Control airports	7	7					Not shown; refer to A/FD
Nonradar Approach/Departure Control airports	7	7					Not shown; refer to A/FD

[1]Includes airspace within 30-nm radius of primary airport, up to 10,000 feet msl

[2]Consult FAR 91.215 for complete regulations, exceptions, and veil airports

[3]See requirements earlier in this chapter.

[4]Lateral limits of Mode C veil shown in thin blue circle surrounding TCA

[5]Inner circle and outer circle, up to 10,000 – 12,000 feet msl

[6]Required if radar service desired

[7]Required if radar service desired. Otherwise, only in non-TCA, non-ARSA ATAs

him. It's too easy to be lulled into a false sense of security when someone on the ground is vectoring you, advising you of other traffic, and generally guiding you through the airspace.

If there's one failure I've noticed among professional instructors, students, and licensed pilots, it's their propensity to keep their eyes either in the cockpit or straight ahead. They apparently haven't been taught to develop a swivel neck and maintain a constant scanning of as much of a 360-degree circle as aircraft structure and physical limitations permit. Back in the dark ages of World War II, this point was continually emphasized during pilot training. You could never be sure what unfriendly fellow was out there with murder in his heart. The best insurance was to see him first, take evasive action, and then fire away.

Now it isn't considered proper to fire at potential targets, but seeing and evading are more than acceptable practices; they are *lifesaving*. If you have your head in the cockpit, for whatever reason, more than 15 seconds in VFR conditions, you ought to be getting a little nervous. That's too long, considering the rate of closure between even light aircraft. Flying in instrument conditions is one thing; attention to the instruments is essential. VFR is something else. "Head up and unlocked" should be the catch phrase of every pilot.

All of which has nothing to do with radio communications—except to observe that in the process of communicating and being monitored on radar, do not allow yourself relaxed alertness. Your safety in flight is enhanced by those on the ground who are directing and advising you, but you still have the responsibility to be your own protector and defender. A swivel neck and sharp eyes are the best shields against disaster in the air.

11
Approach and Departure Control

The book began with radio communications and operations at uncontrolled multicom and unicom airports, followed by tower-only airports. The next airspace element of concern is the Approach/Departure Control facility. Suffice it to say that if one is to limit his flying to only around the smaller fields, with or without a tower, the subject at hand has little significance. For those who want to expand their flight horizons, understanding the role of Approach and Departure and the radio contacts with both is essential.

THE ROLE OF APPROACH AND DEPARTURE CONTROL

Described in one word, the Approach/Departure Control facility is a middleman. In this respect, Departure assumes radar and communications control of a departing aircraft after it has taken off from a tower-controlled airport and might still be in the outer fringes of the ATA. Once released by the tower controller, Departure then guides the aircraft through the 30 to 40 mile airspace for which it is responsible and out into the open country. At that point, it turns the aircraft over to an Air Route Traffic Control Center or terminates radar contact. Conversely, Approach leads arriving aircraft from the outside world into the TCA (Class B) or ARSA (Class C) airspace. When close to the limits of the ATA, it turns the aircraft over to the tower for landing instructions.

As is probably self-evident, the basic role of Approach/Departure is to ensure the orderly flow of traffic into and out of the busier terminals. Without it, the tower controller at higher-volume airports would face an impossible task of vectoring, sequencing, separating, and directing landing and departing aircraft—all within 30 miles or so of the primary as well as satellite airports. It just couldn't be done with any measure of safety. Hence the need for and value of this middleman.

WHERE IS APPROACH/DEPARTURE LOCATED?

What I mean by "Where is Approach/Departure located?" is the physical location of the facility on the airport—not which airports have the service. The location is either in the tower structure of the primary airport or in a nearby radar-equipped building where it can coordinate closely with the tower personnel.

Three acronyms commonly establish the locations: Tracab (terminal radar control in the tower cab); Tracon (terminal radar control); and Rapcon (radar approach control). All provide the same service, but Tracabs are relatively rare, Tracons the most common. The latter, located in the tower structure below the tower cab, is a darkened room, complete with radar, radio, telephone, and computer equipment. A Rapcon is associated with a military Approach/Departure facility that provides radar service to nearby civilian airports and civil aircraft.

To complete the picture, though, the service doesn't always require a facility to be on or near a given airport. Recall, for example, the Clarksburg, West Virginia, illustration in chapter 10. There, when the local Approach/Departure is closed, the Cleveland Center handles arriving and departing traffic through its remoted radar and communications facilities.

Also, I should note that a few airports have nonradar Approach/Departure Control, either on or off the field. Columbia, Missouri, is one example (FIG. 11-1), and

---

COLUMBIA REGIONAL   (COU)   10 SE   UTC−6(−5DT)   38°49'05"N 92°13'10"W   **KANSAS CITY**
    889   B   S4   FUEL 100LL, JET A   OX 2   ARFF Index A   **H-4G, L-21A**
    RWY 02-20: H6500X150 (CONC-GRVD)   S-92, D-125, DT-215   HIRL   **IAP**
      RWY 02: MALSR.   RWY 20: ODALS. VASI(V4L)—GA 3.0° TCH 39'.
    RWY 13-31: H4401X75 (ASPH)   S-12, D-16   MIRL
      RWY 13: REIL. VASI(V2L)—GA 3.0° TCH 44'. Road.   RWY 31: REIL. VASI(V2L)—GA 3.15° TCH 33'.
    AIRPORT REMARKS: Attended Mon–Fri 1200–0600Z‡ and Sat–Sun 1300–0500Z‡. PPR for unscheduled air carrier
      operations with more than 30 passenger seats, call safety officer 314–443–2811. ARFF Index B level svc avbl on
      req. When twr closed ACTIVATE HIRL Rwy 02–20 MALSR Rwy 02 and ODALS Rwy 20—119.3. NOTE: See
      SPECIAL NOTICE–Simultaneous Operations on Intersecting Runways.
    COMMUNICATIONS: CTAF 119.3   ATIS 128.45 (1300–0500Z‡)   UNICOM 122.95
      FSS (COU) on arpt 122.65 122.2 TF 1–800–WX–BRIEF. NOTAM FILE COU.
    ➤ APP/DEP CON 120.0 (1300–0500Z‡)   KANSAS CITY CENTER APP/DEP CON 118.4 (0500-1300Z‡)
      TOWER 119.3 (1300-0500Z‡)   GND CON 121.6
    RADIO AIDS TO NAVIGATION: NOTAM FILE COU. VHF/DF ctc COLUMBIA FSS
      HALLSVILLE (L) VORTAC 114.2   HLV   Chan 89   39°06'49"N 92°07'41"W   188° 18.2 NM to fld. 920/6E
      (L) VOR/DME 110.2   COU   Chan 39   38°48'39"N 92°13'05"W   at fld. 883/3E. HIWAS.
      ZODIA NDB (LOM) 407   CO   38°43'00"N 92°16'06"W   018° 6.5 NM to fld. Unmonitored.
      ILS/DME 110.7 I-COU Chan 44 Rwy 02 LOM ZODIA NDB. ILS Unmonitored when twr clsd.

Fig. 11-1. *Columbia, Missouri, is one example of a nonradar Approach/Departure Control facility located on the airport.*

Great Barrington, Massachusetts, (FIG. 11-2), served by the Albany, New York, Tracon, is another. In these cases, control in the airport vicinity is by "nonradar," or "manual," procedures involving pilot position reports and the controller's memory and notepad. Obviously, this system is not very precise for traffic control in IMC weather, but it's certainly better than no control at all.

GREAT BARRINGTON   (GBR)   2 W   UTC-5(-4DT)   42°11'03"N 73°24'13"W   NEW YORK
   739   B   S4   FUEL 100LL   L-25C, 28H
RWY 11-29: H2579X50 (ASPH)   S-8   LIRL   IAP
   RWY 11: Thld dsplcd 170'. Trees.   RWY 29: VASI(NSTD). Thld dsplcd 75'. Trees.
AIRPORT REMARKS: Attended dalgt hours. Arpt lgts opr dusk-0400Z‡. ACTIVATE LIRL Rwy 11-29 and VASI
   Rwy 29 after 0400Z‡—121.6. For rotating bcn after 0400Z‡ call 413-528-1010. Rwy 11 lgtd thld relocated
   170 ft; 2409 ft of rwy usable for ngt ops.
COMMUNICATIONS: CTAF/UNICOM 122.8
   BURLINGTON FSS (BTV) TF 1-800-WX-BRIEF. NOTAM FILE BTV.
➤ ALBANY APP/DEP CON 125.0
RADIO AIDS TO NAVIGATION: NOTAM FILE BTV.
   CHESTER (L) VOR/DME 115.1   CTR   Chan 98   42°17'28"N 72°56'59"W   266° 21.2 NM to fld. 1600/13W.
   NDB (MHW) 395   GBR   42°10'58"N 73°24'16"W   at fld

Fig. 11-2. *Albany, New York, is responsible for nonradar Approach/Departure Control at Great Barrington, Massachusetts.*

The *A/FD* is the only source to determine whether a given airport has the service. If an ® precedes "APP/DEP CON," radar control is available. If there is no ®, the service is nonradar. And, if you don't see "APP/DEP CON" on the *A/FD* at all, the service doesn't exist.

DEPARTURE FROM A TCA (CLASS B) AIRPORT

To be more specific about all of this, along with examples of the recommended radio phraseology, let's say that you're on the ramp at Kansas City International, which, as you know by now, is a TCA. The engine is started and you're ready to go.

First, tune to the ATIS and remember the information as well as the phonetic designation of the report. Now you make a call to a function that I've previously referred to only in passing: Clearance Delivery (CD)—a function that exists primarily at TCA and ARSA airports.

The purpose of Clearance Delivery is to relieve congestion or excessive communication demands at the Ground Control position. CD has its own frequency, which is published in the *A/FD*. Once contacted, CD's job is to relay to the IFR pilot his IFR clearance. For those going VFR, the instructions consist of the following sequence:

1. Clearance to depart
2. Initial altitude to maintain
3. Departure Control frequency
4. Squawk code

As CD relays this data, be sure to copy it, and then, in an abbreviated form, read it back to the controller so that he knows that you have correctly recorded the instructions. (Suggestion: To make copying easier and faster, write down the main heading in advance and then merely fill in the blanks.)

After listening on the frequency for a few seconds to be certain that the air is clear, the dialogue would go like this:

You: International Clearance, Cherokee One Four Six One Tango.

CD: *Cherokee One Four Six One Tango, International Clearance.*

You: Clearance, Cherokee One Four Six One Tango at the general aviation ramp with India. VFR to Omaha, requesting six thousand five hundred.

CD: *Roger. Cherokee One Four Six One Tango cleared to depart the TCA (or Class B airspace). Maintain four thousand. Departure frequency will be one one niner point zero. Squawk two five two zero.*

You: Understand Cherokee One Four Six One Tango cleared to depart the TCA (or Class B airspace), maintain four thousand, one one niner point zero, two five two zero. Six One Tango.

CD: *Readback correct, Six One Tango. Contact Ground.*

You: Will do. Six One Tango.

Two points: When Clearance Delivery gives you your clearance, that is your permission to operate your aircraft in the TCA. That was it, and no further authority is required. Second, you are given a transponder squawk—2520 in this case. Immediately set your transponder accordingly. Don't wait. You might forget either the code or to set it at all. And turn the transponder switch to STANDBY—not the ON or ALT position. Change to ALT only after you've been cleared for takeoff and are taxiing from the hold line to the active runway.

Now comes the call to Ground Control:

You: International Ground, Cherokee One Four Six One Tango at general aviation ramp with clearance and Information India. Ready to taxi.

CD: *Cherokee One Four Six One Tango, taxi to Runway One Niner.*

You taxi out, complete the pretakeoff check, and move to the runway hold line. Next comes the call to the tower:

You: International Tower, Cherokee One Four Six One Tango ready for takeoff (or "ready to go").

Twr: *Cherokee Six One Tango, Roger. Turn right to two seven zero. Cleared for takeoff.*

You: Roger. Cherokee Six One Tango. [Now turn your transponder to ALT.]

After you're airborne, the tower contacts you again:

Twr: *Cherokee Six One Tango, contact Departure.*

You: Roger. Cherokee Six One Tango. Good day.

Note that the tower said nothing about the frequency on which to contact Departure. Also, you didn't have to request the frequency change. The reason is that all factors relative to your clearance had already been coordinated between Clearance, Tower, and Departure.

If you have two radios, change the frequency in one from Ground Control to Departure just before calling the tower. If you can set up your radios so that you're always one step ahead of the facility you're currently talking to, the transition from one to the next will be simplified. I discussed this in chapter 9, but it's worth reemphasizing as just one more element of good cockpit organization.

Make the next call following the tower's approval to contact Departure:

> **You:**   Kansas City Departure, Cherokee One Four Six One Tango is with you, out of one thousand four hundred for four thousand. [Departure is well aware of your clearance and your squawk code. Consequently, all you need to do is identify your aircraft, your present altitude, and the altitude to which you are climbing. "With you" simply establishes the contact.]

> *Dep:*   *Cherokee Six One Tango, radar contact. Climb and maintain six thousand five hundred. Turn right heading three four zero.*

> **You:**   Roger. Out of one thousand seven hundred for six thousand five hundred. To three five zero. Cherokee Six One Tango.

> *Dep:*   *Cherokee Six One Tango, that heading is three four zero.*

> **You:**   Roger. Three four zero on the heading. Cherokee Six One Tango.

A few minutes later:

> *Dep:*   *Cherokee Six One Tango, position two zero miles north of International, departing the Kansas City TCA. Resume own navigation. Stand by for traffic advisories.* [This means that you are out of the TCA but still in departure control's radar area. He thus wants you to stay with him for possible traffic advisories.]

> **You:**   Roger. Cherokee Six One Tango. Standing by. [You might not hear anything further from Departure until you leave the radar coverage area as follows.]

> *Dep:*   *Cherokee Six One Tango, radar service terminated. Squawk one two zero zero, change to advisory frequency approved.* ["Advisory frequency" is any frequency you now wish—Center, Flight Service, enroute airport towers, etc. There is a chance that radar service will not be terminated and that Departure will "hand you off" to the appropriate Air Route Traffic Control Center for further traffic advisories. I'll get into that in the next chapter.]

> **You:**   Roger. Cherokee Six One Tango. Good day. [Now change the transponder to 1200.]

DEPARTING AN AIRPORT UNDERLYING A TCA

You're departing Spirit of St. Louis, Kansas City Downtown, Peachtree DeKalb (Atlanta), or any other field that lies under but is not in a TCA. What differences, if any, does this situation pose as far as clearances, radio contacts, and the like are concerned?

Assume that you're leaving Spirit, your route of flight is to the northeast, and your desired altitude will put you squarely in the TCA. The sequence of contacts then goes like this:

You: Spirit Ground, Cherokee One Four Six One Tango at the terminal, ready to taxi, VFR northeast bound with Information Delta.

CD: *Cherokee One Four Six One Tango, taxi to Runway Two Six left.*

The runup is complete, you've taxied to the hold line, the transponder is on 1200, the switch in STANDBY:

You: Spirit Tower, Cherokee One Four Six One Tango, ready for takeoff, northeast departure.

Twr: *Cherokee Six One Tango, cleared for takeoff. Northeast departure approved.*

You: Roger, cleared for takeoff, Cherokee Six One Tango.

While taxiing onto the runway, change the transponder switch to ALT and get ready to roll. When you are safely off the ground, the tower should say:

Twr: *Cherokee Six One Tango, contact St. Louis Approach, one two six point seven.*

You: Roger. One two six point seven, Cherokee Six One Tango.

Now, dial in 126.7—if you didn't know the frequency in advance—and call Approach:

You: St. Louis Approach, Cherokee One Four Six One Tango.

App: *Cherokee Six One Tango, St. Louis Approach.*

You: Approach, Cherokee Six One Tango just off Spirit, heading of zero three zero at one thousand eight hundred, requesting seven thousand five hundred to Decatur and clearance to transit the TCA.

App: *Cherokee Six One Tango, squawk zero two five six and ident. Remain outside the TCA until radar contact.*

You: Cherokee Six One Tango squawking zero two five six.

You must contact Approach Control at the primary TCA airport for clearance into the TCA. (Remember why it's Approach and not Departure? If your recollection is faint, go back to chapter 10.) If you do forget when the time comes and happen to call Departure instead of Approach, the controller won't subject you to public reprimands.

Should you say "St. Louis Departure" and the controller acknowledges with "St. Louis Approach," go ahead with what you wanted to say and henceforth refer to the service as "Approach."

An important point here: The fact that Approach has given you a discrete squawk and asked you to ident does not constitute clearance into the TCA. Whatever you do, don't plunge merrily along. Level off to avoid busting the 2,000-foot floor; circle, do S-turns, or anything else that will keep you out of the sacred area until you hear something like this:

App: *Cherokee Six One Tango, radar contact. Cleared into the TCA. Turn right, heading zero four zero. Climb and maintain three thousand five hundred.*

You: Roger, understand cleared into the TCA. Right to zero four zero. Out of one thousand eight hundred for three thousand five hundred. Cherokee Six One Tango.

A word about altitude reporting: When working with Departure or Approach, always advise when you're leaving one altitude for a newly assigned altitude. The only time you don't have to worry about such reports is when you are advised to maintain VFR altitudes at your discretion. "At your discretion" means that you can climb or descend as you desire. In a controlled environment, however, you must report altitude changes and maintain the last one assigned. Unless specifically requested by the controller, you are not required to report reaching your new altitude, although some controllers prefer that you do. If for any reason you can't stay VFR at the assigned altitude, advise Approach or Departure and request a new altitude. Jumping around horizontally or vertically on your own in a TCA is very, very *verboten*. Unless you hear "at your discretion," stick to the altitude and heading assigned.

To continue the illustration:

You: Approach, Cherokee Six One Tango level at three thousand five hundred.

App: *Cherokee Six One Tango, Roger. Traffic at twelve o'clock, three miles southeast bound at three thousand.*

You: Negative contact. Cherokee Six One Tango.

Perhaps a couple of minutes later:

App: *Cherokee Six One Tango, traffic no longer a factor. Climb and maintain seven thousand five hundred.*

You: Roger, out of three thousand five hundred for seven thousand five hundred. Cherokee Six One Tango.

When at the assigned altitude:

You: Approach, Cherokee Six One Tango level at seven thousand five hundred.

App: *Cherokee Six One Tango. Roger.*

When you are clear of the TCA, Approach will advise you accordingly:

App: *Cherokee Six One Tango, position two zero miles northeast of St. Louis, departing the TCA. Squawk one two zero zero. Radar service terminated. Frequency change approved. Resume own navigation.*

You: Roger. Cherokee Six One Tango. Thanks for your help. Good day.

DEPARTING AN ARSA (CLASS C) AIRPORT

An ARSA departure offers two possible advantages for the VFR pilot over a TCA. First, if you have the traffic ahead of you (or the traffic you are supposed to follow) in sight and can maintain visual separation, Departure might clear you to turn on course and climb to your cruising altitude. In other words, once that clearance is issued, your departure is much like departing a typical, nonradar, tower-controlled airport. There would be minimum vectoring, if any, and ATC would contact you only to give you advisories of other traffic that might be in your vicinity. The principle of visual separation for VFR aircraft would be in operation. All of this means, of course, that your flight would be more direct and the departure from the ARSA more rapid. This might not always be the standard procedure. Much depends on weather conditions, visibility, and the volume of traffic in the ARSA. It is, however, a prerogative of ATC, when circumstances permit.

A second advantage for the VFR pilot who wants to get up and go is the size of the ARSA. The ARSA has only a 10-mile radius to the limits of the outer circle, compared to 15 to 30 miles for a TCA. Consequently, you clear the regulated airspace much more rapidly and are free to fly on your own or contact Center, if you wish.

Clearance Delivery

Most ARSAs, but not all, have Clearance Delivery (the frequency is in the *Airport/Facility Directory*). If no Clearance Delivery exists, merely contact Ground Control and communicate your intentions. Ground provides the same service as Clearance Delivery and then clears you to taxi.

Whichever the case, the radio calls are identical. Listen to the ATIS, and then tune to the Clearance frequency. To illustrate the communication one more time, let's say you're leaving the Birmingham, Alabama, ARSA for Memphis:

You: Birmingham Clearance, Cherokee One Four Six One Tango at Hangar One with Echo. VFR Memphis via Victor 159, requesting six thousand five hundred.

CD: *Cherokee One Four Six One Tango, Birmingham Clearance, Cleared to depart the Birmingham ARSA. Maintain three thousand. Departure frequency one two four point five, squawk two six three six.*

You: Roger, cleared to depart the ARSA, three thousand, one two four point five, and two six three six. Cherokee Six One Tango.

CD: *Cherokee Six One Tango, readback correct. Contact Ground.*

Next, the usual call to Ground and then to the tower.

Tower and Departure Control.

You're at the hold line and ready to go:

You: Birmingham Tower, Cherokee One Four Six One Tango ready for takeoff.

Twr: *Cherokee Six One Tango, cleared for takeoff. Fly runway heading.*

In this transmission, the tower might instruct you to "contact Departure" when airborne. If not, be sure to request the frequency change after you are off the ground and have the aircraft under control. Don't go from one frequency to another in a controlled area without permission.

The call to Departure is much the same as in a TCA:

You: Birmingham Departure, Cherokee One Four Six One Tango is with you, out of one thousand two hundred for three thousand.

Dep: *Cherokee Six One Tango, radar contact. Turn right heading three one zero. Report reaching three thousand.*

You: Right to three one zero and report three thousand. Cherokee Six One Tango.

A few minutes later:

You: Birmingham Departure, Cherokee Six One Tango level at three thousand.

Dep: *Cherokee Six One Tango, Roger. Climb and maintain six thousand five hundred.*

You: Roger, out of three thousand for six thousand five hundred. Cherokee Six One Tango.

When at altitude:

You: Birmingham Departure, Cherokee Six One Tango level at six thousand five hundred.

Dep: *Cherokee Six One Tango, Roger.*

When you pass the 10-mile outer circle, you can request termination of radar advisories, if you desire. Otherwise, radar service continues until you reach the 20-mile perimeter of the outer area, when you hear the final word from Departure:

Dep: *Cherokee Six One Tango, position twenty miles northwest of Birmingham, radar service terminated. Squawk one two zero zero. Change to advisory frequency approved.*

You: Roger, Cherokee Six One Tango. Good day. [Be sure to change the transponder to 1200.]

Between these exchanges might come vectors, advisories, or even safety alerts, depending on conditions or traffic with the ARSA. Or, as mentioned earlier, if the visibility is good and the traffic light, Departure Control has the prerogative to authorize a departing VFR aircraft to proceed directly on course and to continue its climb to the desired cruising altitude. That does not obviate the need for the pilot to maintain radio contact with Departure until he is clear of at least the outer circle's 10-mile radius. In the outer area, which is not part of the ARSA, radar service can be declined by the pilot but not by ATC.

Departing a Satellite Airport within the Inner Circle

If you're departing a no-tower airport that is within the ATA, say, four miles from the primary airport, and you can't reach the primary airport's tower by radio, you can still take off, but you must follow the established traffic pattern for the satellite. You must contact the tower as soon as possible after takeoff.

DEPARTURE FROM OTHER RADAR AIRPORTS

This time you're leaving from an airport such as Springfield, Missouri. Like Clarksburg in chapter 10, there are quite a few Springfields around that have no TCA or ARSA but offer radar Approach and Departure Control nonetheless. You don't have to use this service, but why not?

To avoid repetition, let's assume that you have the ATIS, have been cleared by Ground to the hold area, and are ready to contact the tower:

You: Springfield Tower, Cherokee One Four Six One Tango ready for takeoff, north departure.

Twr: Cherokee Six One Tango, cleared for takeoff. North departure approved.

You're now off the ground:

You: Springfield Tower, Cherokee Six One Tango requests frequency change to Departure.

Twr: Cherokee Six One Tango, frequency change approved.

So you thought the tower would automatically give you the Departure frequency? Wrong. You should have done your homework and determined the frequency before you got in the airplane. That's just one small element of preflight preparation—preparations that the amateur neglects and the professional doesn't forget. But you've screwed up a little, so:

You: Tower, Cherokee Six One Tango. What is the Departure frequency, north?

Twr: Cherokee Six One Tango, contact Departure on one two four point niner five.

You: One two four point niner five. Cherokee Six One Tango. Thank you.

You: Springfield Departure, Cherokee One Four Six One Tango is four north of Springfield, out of two thousand for six thousand five hundred, VFR Kansas City, squawking one two zero zero. Request traffic advisories.

Dep: *Cherokee Six One Tango, squawk zero two zero zero and ident.*

You: Cherokee Six One Tango squawking zero two zero zero.

Dep: *Cherokee Six One Tango, radar contact. Report reaching six thousand five hundred.*

From here on, Departure might vector you, alert you to other aircraft, or do whatever else is necessary to clear you from the area. When you reach your altitude, however, don't forget to communicate that fact:

You: Springfield Departure, Cherokee Six One Tango. Level at six thousand five hundred.

Dep: *Cherokee Six One Tango, Roger.*

Eventually, you'll hear this:

Dep: *Cherokee Six One Tango, radar service terminated. Squawk one two zero zero. Change to advisory frequency approved.*

You: Roger. Cherokee Six One Tango. Good day.

Now you're free to turn to any radio frequency that you want—Flight Service, Flight Watch, unicom of any airports you'll be passing over, a tower or ATIS of a larger airport on your route, etc. The choice is yours.

Conclusion

That fairly well sums up the Departure Control procedures. As is hopefully apparent, there's nothing complicated about the process or what's required of the pilot. If you've done just a little homework ahead of time, if you know the frequencies, know what to say to what facility, and have a reasonably good idea of what you can expect to hear, getting out of even the busiest TCA (Class B) is not difficult. Beyond those basics, once contact is made with each facility—Clearance, Ground, Tower, and Departure—it's really no more than a matter of listening, following instructions, and paying attention to what's going on outside the cockpit, none of which should be a challenge to any pilot.

But now that you're out of the controlled airport environment, the next matter of concern is how to get back into one. So with that in mind, let's turn to Approach Control and its role in the scheme of things.

APPROACH TO A TCA AIRPORT

The various radio procedures should be fairly well in your mind by now, and those required for entering a TCA are only minor variations of several that I have already illustrated. Regardless, I might as well start at the beginning just to be sure that the entire range of communications has been covered.

Before calling on Approach for radar assistance into any airport, you have to be prepared to supply certain information. That information can be easily remembered if you use the acronym IPAI/DS ("eye-paids"):

I = Aircraft *Identification*—type and full N-number
P = *Position*—where you are geographically
A = Present *Altitude*
I/D = *Intentions* and/or *destination*
S = *Squawk*—what code you are currently squawking.

The reason for "*Intentions* and/or *Destination*" is that the two might not be the same. For instance, you're approaching the Dallas/Ft. Worth TCA from the east en-route to Abilene, Texas. You merely want (or intend) to cross through the TCA (Class B airspace), and the fact that Abilene is your ultimate destination is of little importance to Dallas Approach. It's not wrong to include "Abilene" in the radio call, but the main thrust of the call is your *intention* to transit the TCA, not your final destination. In reality, the I/D portion of the acronym can be stated one of three ways:

1. "Request clearance through the TCA (Class B airspace) to the west"
2. "Request clearance through the TCA (Class B airspace) to the west to Abilene"
3. "Request clearance through the TCA (Class B airspace) to Abilene."

Of course, if you intend to land anywhere in the TCA, as Ft. Worth, or in the TCA's immediate area, that's your destination, and the name of the airport is what Approach needs to know. The I/D portion of the call would then be:

"Landing Ft. Worth Meacham."

Keeping this sequence in mind, and rehearsing what you're going to say before getting on the mike, conveys an image of competence—which, in turn, increases the likelihood of receiving the assistance that you want. Controllers like to deal with pilots who sound as though they know what they're doing.

Let's use St. Louis as an example again. You're coming in from the west and want to land at Lambert Field, a TCA airport. You're over the Foristell VOR, which is about 30 statute miles west of Lambert and about 10 miles from the outermost TCA ring. Now is the time to check the ATIS and call Approach:

You: St. Louis Approach, Cherokee One Four Six One Tango.

App: *Cherokee One Four Six One Tango, St. Louis Approach.*

You: Cherokee One Four Six One Tango, over Foristell VOR at seven tho' hundred, landing Lambert. Squawking one two zero zero with I Lima.

App: *Cherokee Six One Tango, squawk zero two three five and ident. Ᵽᵉᵐ... outside the TCA until radar contact.*

You: Cherokee Six One Tango squawking zero two three five.

Remember that just being given a squawk code does not authorize you to enter the TCA nor does the controller's statement that radar contact has been established. It's only when you hear "cleared into the TCA (or ". . . into the St. Louis Class B airspace") that you can penetrate that airspace. "Cleared" is the key word. (Sorry to beat this point, but the FAA is very emphatic about it.)

App: *Cherokee Six One Tango, radar contact. Cleared into the TCA (Class B airspace). Descend and maintain four thousand five hundred.*

You: Roger, understand cleared into the TCA. Out of seven thousand five hundred for four thousand five hundred, Cherokee Six One Tango.

A few minutes later:

You: Approach, Cherokee Six One Tango level at four thousand five hundred.

App: *Cherokee Six One Tango. Roger. Turn left, heading zero eight five.*

You: Left to zero eight five, Cherokee Six One Tango.

From this point on, Approach might give various instructions relative to new headings, altitude changes, advisories of other traffic in or near your line of flight, and the like. Each instruction or advisory should be promptly acknowledged—and not just with "Roger." Remember to repeat tersely whatever the instruction is or to advise Approach if you see or don't see the traffic that is in your vicinity.

Eventually, after you've been vectored and granted permission to descend further, Approach turns you over to the tower:

App: *Cherokee Six One Tango, contact St. Louis Tower on one one eight point five.*

You: Roger. One one eight point five. Cherokee Six One Tango.

Now what do you tell the tower? Almost nothing. The controller knows you're coming because Approach has so advised him. Thus there's no need for a time-consuming position/altitude/squawk report. Just this:

You: St. Louis Tower, Cherokee One Four Six One Tango is with you, level at two thousand.

Twr: *Cherokee Six One Tango. Enter left base for Runway Three Zero Left.*

You: Roger, left base for Three Zero Left. Cherokee Six One Tango.

The instructions and landing continue in the normal pattern. To repeat some principles:

- Never leave an assigned altitude in a TCA without permission.

- Always report when leaving one altitude for another that has been newly assigned. Report reaching the new altitude if you are instructed to do so (otherwise this is optional).

- "At pilot's discretion" means to climb or descend when you wish.

- When told to "ident," you don't need to acknowledge with "Cherokee Six One Tango identing." Just push the button. Do not push the button unless you are instructed to ident.

- Don't change from an assigned squawk back to 1200 until you are told.

- If, for any reason, you can't remain at the assigned altitude because of weather, request permission to climb or descend:

You: Approach, Cherokee Six One Tango requests lower account ceilings.

App: *Cherokee Six One Tango, descend and maintain two thousand five hundred.*

You: Leaving three thousand five hundred for two thousand five hundred, Cherokee Six One Tango.

One other point is appropriate to emphasize in this discussion: When Approach, Departure, or tower gives you an instruction, obey it immediately. For example, Approach tells you to turn to a heading of 120 degrees. As soon as you hear the heading, begin your turn. Don't keep flying on your present course while you pick up the mike (perhaps having to wait until the air is clear) to acknowledge the instruction. Do it now. Then, while in the turn, call Approach and confirm the 120-degree heading. Similarly, if you're told to climb or descend, obey immediately and confirm that action as soon as you can. The controller probably has a good reason for varying your line of flight, so don't hesitate to do what you're told.

APPROACHING AN AIRPORT THAT UNDERLIES A TCA

Instead of landing at Lambert Field, you want to go into Spirit of St. Louis—which underlies Lambert's TCA to the southwest. This time, however, you're coming in from the northeast and would like vectors through the TCA rather than detouring around it. Despite the fact that you're landing at a non-TCA airport, the radio communications are much the same.

To set the scene, you're cruising at 6,500 feet and homing on the St. Louis VORTAC, which is about 10 statute miles (sm) northwest of Lambert. Well outside the

TCA, make your first call:

You: St. Louis Approach, Cherokee One Four Six One Tango.

App: *Cherokee One Four Six One Tango, St. Louis Approach.*

You: Approach, Cherokee One Four Six One Tango is over Bunker Hill on the St. Louis zero five niner radial, level at six thousand five hundred, landing Spirit. Squawking one two zero with Information Papa.

App: *Cherokee Six One Tango, squawk zero two five two and ident. Remain outside the TCA until radar contact.*

You: Cherokee Six One Tango, zero two five two.

App: *Cherokee Six One Tango, radar contact three zero miles east of the St. Louis VOR. Cleared into the TCA. Descend and maintain four thousand five hundred. Turn left, heading one niner zero.*

You: Roger, understand cleared into the TCA. Out of six thousand five hundred for four thousand five hundred, left to one niner zero. Cherokee Six One Tango.

A little later:

App: *Cherokee Six One Tango, turn right, heading two seven zero.*

You: Right to two seven zero, Cherokee One Tango.

App: *Cherokee Six One Tango, descend and maintain two thousand five hundred.*

You: Out of four thousand five hundred for two thousand five hundred. Cherokee Six One Tango.

As you near the Spirit Airport, you'll be below the 3,000-foot TCA floor. About this time, Approach will call you:

App: Cherokee Six One Tango, position five miles east of Spirit. Contact Spirit Tower, One Two Four point Seven Five.

You: One Two Four point Seven Five. Thank you for your help. Cherokee Six One Tango.

Now contact the Spirit Tower for pattern and landing instructions: "Spirit Tower, Cherokee One Four Six One Tango is with you 5 miles east, level at two thousand five hundred."

Using the same situation of an airport underlying a TCA, let's suppose that you are not familiar with the airport and there is no VOR or nondirectional beacon (NDB) on the field for guidance. In other words, you need assistance.

You're again coming into the St. Louis TCA from the northeast. After radar contact has been established and you are cleared into the TCA, Approach calls you: "Cherokee Six One Tango, cleared on course [meaning "resume your own navigation]. Descend and maintain two thousand five hundred."

Being uncertain of the location of the airport, you request further help:

You: Roger. Cherokee Six One Tango is out of three thousand five hundred for two thousand five hundred. Request vectors to Spirit.

App: *Cherokee Six One Tango, Roger. Turn right, heading two seven zero to Spirit.*

You: Right to two seven zero. Cherokee Six One Tango.

App: *Cherokee Six One Tango, Spirit is at twelve o'clock, eighteen miles. Report when you have it in sight.*

You: Will do, Cherokee Six One Tango.

In a few minutes:

You: Approach, Cherokee Six One Tango has Spirit in sight.

Approach then turns you over to Spirit Tower as before.

Another situation isn't unusual when the TCA is busy and your destination is an underlying airport: permission to enter the TCA is not granted. In that case, getting into a field such as Spirit from the northeast means skirting the TCA. It adds time and mileage, but you have no alternative:

You: St. Louis Approach, Cherokee One Four Six One Tango.

App: *Cherokee One Four Six One Tango, St. Louis Approach.*

You: Approach, Cherokee Six One Tango is over Bunker Hill at six thousand five hundred, squawking one two zero zero with Information Papa. Request clearance into the TCA (Class B airspace) for landing Spirit.

App: *Cherokee Six One Tango, unable. Change to Spirit frequency one two four point seven five and remain clear of the TCA (Class B airspace).*

You: Roger, Approach. Cherokee Six One Tango remaining clear.

With those instructions, your only alternative is to detour around the airspace and stay beneath all floors until the Spirit Airport is in sight. At that point, make the routine call to Spirit's Tower.

TRANSITING A TCA

This time, you don't want to land in or under the TCA but instead pass through it to a more distant destination. What you say and what you do are almost identical to landing procedures. Assume that you're east of St. Louis and are heading for Jefferson City, Missouri. The straight line from your present position would put you right through the TCA. You could go around the whole thing, but, again, that would only add time and fuel costs—so why not take the economical way out? That being the more intelligent avenue, you plan what you're going to say and call Approach:

You: St. Louis Approach, Cherokee One Four Six One Tango.

App: *Cherokee One Four Six One Tango, St. Louis Approach.*

You: Approach, Cherokee One Four Six One Tango approaching Troy VOR on zero seven six radial, level at six thousand five hundred to Jeff City. Squawking one two zero zero. Request vectors through the TCA.

App: *Cherokee Six One Tango, squawk zero two six four and ident. Remain outside the TCA until radar contact.*

You: Cherokee Six One Tango squawking zero two six four, and remaining clear.

App: *Cherokee Six One Tango, radar contact. Cleared through the TCA. Maintain six thousand five hundred. Turn right, heading two seven zero.*

A pause here to explain something: you reported your position as "approaching Troy VOR on zero seven six radial." Why the "076 radial?" Approach knows, of course, where the Troy VOR is located, so it should be easy to spot your aircraft once you've squawked the discrete code. Disregarding that, however, give the radial *from* the VOR in a position report. That makes identification of your aircraft much easier for the controller. "Approaching on the zero seven six radial" alerts him to look for you on the east side of his screen. In this case, the 076 radial also happens to be the centerline of Victor 12, a VOR airway. So you could alternatively report "five miles east of Troy on Victor Twelve."

If, at any time, you are asked by Center or Approach to advise what *radial* you are flying, report it in terms of the *outbound bearing* from the station. If asked your *heading*, report the actual reading on your compass or directional gyro.

Now back to the example:

You: Roger, understand cleared through the TCA. Maintain six thousand five hundred, right to two seven zero. Cherokee Six One Tango.

A few minutes later:

App: *Cherokee Six One Tango, turn left, heading two four five.*

You: Left to two four five. Cherokee Six One Tango.

App: *Cherokee Six One Tango, traffic at eleven o'clock, two miles northeast bound at five thousand five hundred.*

You: Roger, negative contact. Cherokee Six One Tango.

With your eyes scanning the skies at about the eleven o'clock position (don't forget to compensate for any crab angle), you spot the traffic approaching your position, but below you:

You: Approach, Cherokee Six One Tango has the traffic.

App: *Cherokee Six One Tango, Roger.*

Eventually, after whatever vectoring Approach deems necessary to provide the proper aircraft separation, you are clear of the TCA on the west side:

App: *Cherokee Six One Tango, position two zero miles southwest of St. Louis VOR. Departing the TCA. Squawk one two zero zero. Radar service terminated. Frequency change approved.*

You: Roger. One two zero zero—and thank you for your help. Cherokee Six One Tango.

You're on your own again, with Jeff City off in the distance.

APPROACH CONTROL AND THE ARSA

The mandatory Approach Control radar service at an ARSA requires no new or additional radio phraseology. The differences between TCAs and ARSAs are confined to procedures or regulations meaning:

TCAs: Clearance into TCA is required; clearance can be denied.

ARSAs: Radio contact with Approach must be established before entering the ARSA, but a literal "clearance" into the ARSA is not required; use of radar service is mandatory.

If you keep IPAIDS in mind and have the basic TCA (Class B) call pattern mastered (plus, of course, two-way radio communications capability and Mode C), you have all you need to operate in an ARCs (Class C) airspace.

To follow through with an example: You're approaching the Tucson, Arizona, ARSA from the northwest on Victor Airway 308 for landing at Tucson International. About 20 miles out, you contact Approach. Remember that you can enter an ARSA's (Class C) outer area without radio contact but not the 10-mile radius outer circle.

You: Tucson Approach, Cherokee One Four Six One Tango.

App: *Cherokee One Four Six One Tango, Tucson Approach.*

You: Approach, Cherokee Six One Tango is twenty northwest on Victor 303, level at seven thousand five hundred for landing International, squawking one two zero zero with Information Echo.

App: *Cherokee Six One Tango, squawk four five four three and ident.*

You: Four five four three. Cherokee Six One Tango.

App: *Cherokee Six One Tango, radar contact. Fly present heading and descend and maintain six thousand.*

You: Roger, Approach, Six One Tango leaving seven thousand five hundred for six thousand.

Along with altitude changes and possible, or probable, vectors, you finally hear:

App: *Cherokee One Four Six One Tango, contact International Tower one one eight point three.*

You: Roger, one one eight point three, Six One Tango.

You: International Tower, Cherokee One Four Six One Tango is with you, level at four thousand five hundred. (Tucson's elevation is 2,641 feet msl, with a light aircraft pattern altitude of 3,400 feet msl.)

Twr: Roger, Cherokee One Four Six One Tango, enter left base for Runway Two Niner.

You: Left for Two Niner. Six One Tango.

And so it goes until you're down and parked.

Suppose, however, in response to your initial call and position report, that Approach came back with this:

App: Cherokee One Four Six One Tango, stand by.

What do you do now? Nothing. Stay right on course into the airspace. You have established the required communications, which is all that's necessary. Unless the controller: 1) doesn't answer you at all; 2) replies with "Aircraft calling Tucson Approach, stand by"; or 3) specifically calls you by your aircraft N-number and tells you to "remain clear (or outside) of the ARSA (Class C airspace)," you can legally enter the area. Therein lies a major TCA/Class B versus ARSA/Class C difference.

TRANSITING AN ARSA

If you're on a cross-country at normal VFR cruising altitudes in the 6,500-to 8,500-foot range, you won't have much trouble staying above most ARSAs. Their typical 4,000-foot-plus ceiling makes overflying relatively easy—unless, of course, weather forces a lower altitude. In such instances, transiting an ARSA that lies along your route might be necessary to avoid a flight-prolonging detour.

Fortunately, the radio contact is almost identical to that when transiting a TCA. To illustrate, assume that you're enroute from Memphis to the Peachtree Dekalb Airport near Atlanta, and the cloud cover has forced you to cruise at 3,500 feet msl. Nearing the Birmingham, Alabama, ARSA, you make the initial call:

You: Birmingham Approach, Cherokee One Four Six One Tango.

App: Cherokee One Four Six One Tango, Birmingham Approach.

You: Approach, Cherokee One Four Six One Tango is twenty northwest on Victor One Five Niner, level at three thousand five hundred, enroute Peachtree Dekalb, squawking one two zero zero. Request advisories through the ARSA.

App: Roger, Cherokee Six One Tango. Squawk four two five three and ident.

You: Cherokee Six One Tango squawking four two five three.

App: Cherokee Six One Tango, radar contact. Birmingham altimeter is Two Niner Four Seven.

You: Roger, Two Niner Four Seven. Cherokee Six One Tango.

What happens for the next few minutes depends on the volume of traffic in the ARSA and the advisories or safety alerts ATC issues you. But what would you do if ATC, for good reasons, asks you to leave your present altitude and climb to one that would put you in or close to the base of the overcast ceiling? You'd then be less than 500 feet below the cloud layer and in violation of VFR regulations.

The simple answer, and the point for reemphasis is: Don't blindly follow orders that would result in a violation. Advise ATC of the situation. The FARs make it very clear that you're in command of the aircraft. Yes, directions should be obeyed, but ATC might not be aware that an altitude change would make you a rule-breaker. Consequently, if any instruction would cause you to violate a regulation, regardless of ARSAs, TCAs, or otherwise, make that fact known to the controlling agency.

For altitude separation purposes between a VFR and an IFR aircraft, you might be told to climb or descend to a non-VFR cruising altitude, such as 3,000 or 4,000 feet. When the altitude separation is no longer needed, and especially when leaving the regulated airspace, ATC will advise you to "Resume appropriate VFR altitudes." That's your directive to climb or descend to the odd- or even plus-500-feet VFR cruising altitudes dictated by FAR 91.159.

Returning to the Birmingham ARSA exchanges, when you depart the outer area on your way to Atlanta, you'll eventually hear:

App: *Cherokee Six One Tango, position twenty miles east of Birmingham, radar service terminated. Squawk, one two zero zero. Change to advisory frequencies approved.*

You: Roger, Cherokee Six One Tango, good day. [Now change the transponder to 1200.]

APPROACH CONTROL AT CLASS D TOWER-ONLY AIRPORTS

At those Class D airports that have only a tower but no Class B TCA or Class C ARSA, you have the option to use or not use Approach Control. The Clarksburg example described in chapter 10 falls in this category, as do the few so-called "Designated" airports. One situation in which Approach at such airports can be helpful (other than providing advisories, of course) is when you're entering unfamiliar territory and would like directions (vectors) to the airport area. Using Clarksburg again, the following simulates the radio communications for an approach and landing at the city's Benedum Airport:

You: Clarksburg Approach, Cherokee One Four Six One Tango.

App: *Cherokee One Four Six One Tango, Clarksburg Approach.*

You: Approach, Cherokee One Four Six One Tango is over New Martinsville level at five thousand five hundred, landing Benedum. Squawking one two zero zero with Foxtrot.

App: *Cherokee Six One Tango, squawk four five five three and ident.*

You: Four five five three, Six One Tango. Request.

App: *Cherokee Six One Tango go ahead.*

You: Am unfamiliar with the area, Approach, and request vectors to the airport.

App: *Roger, Six One Tango. Radar contact 30 miles northwest. Fly heading one three zero, vectors later to Benedum.*

You: Roger, one three zero. Six One Tango.

Later:

App: *Cherokee One Four Six One Tango, turn left, heading one two zero. Descend pilot's discretion and maintain three thousand.*

You: Left to one two zero, and leaving five thousand and five for three thousand. Cherokee Six One Tango.

You: Approach, Cherokee Six One Tango level at three thousand.

App: *Roger, Six One Tango. Benedum is at your one o'clock position, seven miles. Report the airport in sight.*

You: Approach, Six One Tango has the airport.

App: *Roger, Six One Tango. Contact Benedum Tower on one two six point seven.*

You: One two six point seven. Thank you. Cherokee Six One Tango.

You: Benedum Tower, Cherokee One Four Six One Tango is with you, level at three thousand.

Twr: *Cherokee One Four Six One Tango, enter right base for landing Runway Two One.*

You: Right for Two One, Cherokee Six One Tango.

A few minutes later:

Twr: *Cherokee Six One Tango, cleared to land Two One.*

You: Cleared to land. Six one Tango.

CONCLUSION

The only challenge that Departure and Approach Control pose is the challenge of the unknown. When a pilot feels insecure or uncertain about what to do or say, he takes any action he can to skirt a TCA or an ARSA. As a result, he won't go places he'd like to go. He adds needless miles to an otherwise straight-line cross-country. He lands at an out-of-the-way airport and hopes that some form of transportation is available to get him to his ground destination. He bounces around at 2,500 feet to stay under a TCA when he could have a smooth flight at 4,500 feet in the TCA. And so on. . . . All because contacting Approach or Departure scares him. It is, he feels, beyond his expertise, beyond his scope of knowledge.

This need not be the case. By way of repetition, using Approach and Departure simply means knowing what you're going to say and knowing what you can expect to hear. Once you have the "IPAIDS" down, the rest is just a matter of listening, acknowledging, and obeying. The controller tells you what to do. Obedience within the limits of safety and flight rules and keeping your eyes open are your responsibility.

"But what if I don't understand the instructions? What if they use a term I've never heard of?"

That could—and does—happen. If you don't understand an instruction or a term, ask for clarification: "Say again, please." "I didn't understand." "Am not familiar with that reporting point." "Am not familiar with the area." *Don't just Roger the instruction and then hope that things will work out.* They might not, and upon landing, you might find that the tower controller wants to talk with you.

No one in Approach or Departure is going to crawl through your mike cord if you sound as though you know what you're doing and then ask that an instruction be repeated or clarified. The controller wants to know if you don't know, and he wants to know when you have no uncertainty about what you are to do. That's his job—and that's one reason why instructions are repeated back. Controllers do everything in their power to help you, guide you, and lead you safely to wherever you're going. It's up to you to give them that chance.

So use the services that you're urged to use as a VFR pilot. Know what you're going to say; say it with confidence; say it tersely but distinctly; say it with the ring of a professional. As so many pilots have learned, you'll get the help you need.

12
Air Route
Traffic Control
Centers

Most of the previous chapters have kept you in the general vicinity of the airport—Tower, Approach, Departure, and the like. Now is the time to stray beyond the local aerodrome and consider the pilot's good cross-country friend: The Air Route Traffic Control Center (ARTCC, or "Center," for short).

CENTER'S ROLE IN THE SCHEME OF THINGS

Its name describes its purpose and role: An ARTCC is indeed the *center* for control of all enroute IFR and, when its workload permits, advisories to VFR traffic, within its assigned geographical area. An ARTCC thus ensures the vertical and horizontal separation of IFR aircraft and alerts participating VFR aircraft to potential traffic hazards. In so doing, *control* is the important word. That's Center's job—which means that functions not directly related to the control of traffic, such as filing or amending a flight plan, requesting enroute weather, or asking for weather forecasts, should be directed to the nearest Flight Service Station.

This is not to say that a Center's responsibility is limited solely to enroute traffic control. Although it is not a forecasting agency, it alerts pilots to severe weather that might lie just ahead; in a distress or urgency situation or when a pilot is lost and needs guidance to the nearest field, it can be a life-saver. And, of course, it assumes the Approach Control function at airports when the local Approach facility is closed or at other fields within its radar range that have no such facility at all.

Basically, though, Center is the bridge between Departure Control at the flight's airport of origin and Approach Control at point of landing. Once the aircraft is clear of the TCA (Class B) or ARSA (Class C), Departure drops out of the picture and Center steps in. For the duration of the flight, one or more Centers (depending on the distance traveled) become the enroute controllers, assigning altitudes and vectors to IFR aircraft, advising participating VFR pilots of potential traffic, and generally monitoring the safety of the airways. As the flight nears its destination, Center either terminates the radar service (as it might for VFR aircraft) or hands the pilot off to Approach for sequencing and separation prior to the actual tower-controlled landing.

So, once again, the basic flow from taxi-out to landing is as follows:

1. Ground Control
2. Tower
3. Center
4. Approach Control
5. Tower
6. Ground Control

THE CENTERS AROUND THE COUNTRY

Twenty Air Route Traffic Control Centers are responsible for the enroute traffic in the contiguous 48 states, plus one each in Alaska, Hawaii, Puerto Rico, and Guam. By location, those in the 48 are:

Albuquerque	Houston	Minneapolis
Atlanta	Indianapolis	New York
Boston	Jacksonville	Oakland
Chicago	Kansas City	Salt Lake City
Cleveland	Los Angeles	Seattle
Denver	Memphis	Washington, DC
Fort Worth	Miami	

Geographically, the areas the various Centers control are illustrated in FIG. 12-1.

Fair enough, but what about the geography between almost any two of these locations? Take Washington, DC, and Jacksonville, for example. Referring to FIG. 12-1 again, you'll see that the Washington Center (actually located at Leesburg, Virginia) controls traffic from just north of Washington through most of Virginia, down to the border between North and South Carolina. At that point, the Jacksonville Center takes over. But the two Centers are about 650 miles apart, so with that spread, control can be maintained only through a network of remote communications that tie the more distant radar sites back to the physical location of each Center by microwave links or land lines.

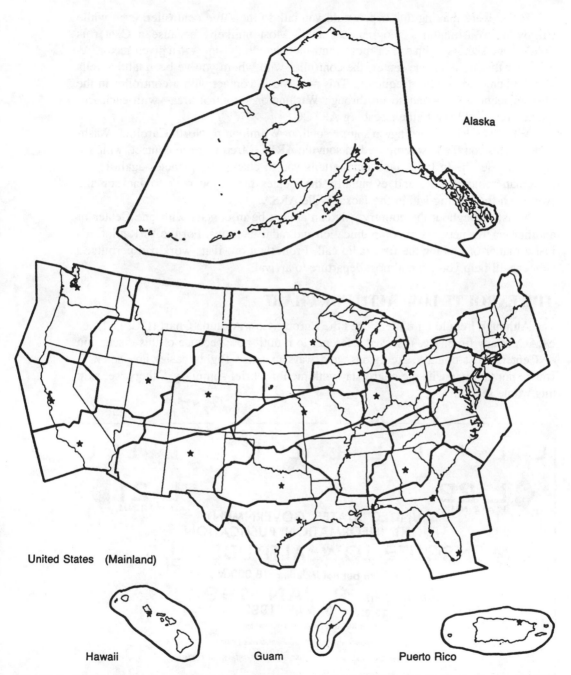

Alaska

United States (Mainland)

Hawaii   Guam   Puerto Rico

Fig. 12-1. *Outline of the locations and geographical areas for which the 24 Air Route Traffic Control Centers are responsible.*

If you were making this trip, would you talk to the same controller, even while still within Washington's geographical area? Most unlikely, because a Center is divided into sectors, with one or more controllers handling flights in a given sector. As you near the limits of one sector, the controller with whom you've been talking tells you to change to another frequency. This puts you in contact with a controller in the next adjacent sector—and so on through Washington's control area—with each frequency remoted back to the Leesburg ARTCC.

When you leave Washington's area, south of Wilmington, North Carolina, Washington gives you the first remoted Jacksonville ARTCC frequency to contact, which is at Myrtle Beach. In time, Jacksonville tells you to change frequencies again to the Savannah remoted site, and eventually, the Center turns you over to Jacksonville Approach for landing within the Jacksonville ARSA.

Thus, throughout the country, you can always be in contact with one Center or another via this network of communication and radar facilities. For the VFR pilot, it's just a matter of making the first radio call. From then on, if its workload permits, a Center will help you from almost departure to arrival.

THE ENROUTE LOW ALTITUDE CHART

Although I could have mentioned the Enroute Low Altitude Chart (FIG. 12-2), or ELAC, at any time, I've reserved reference to it until now because of this discussion of Center, cross-country excursions, and ways to make flights easier and safer. If you're not familiar with the chart, stay with me for a brief summary. Otherwise, skip this whole section.

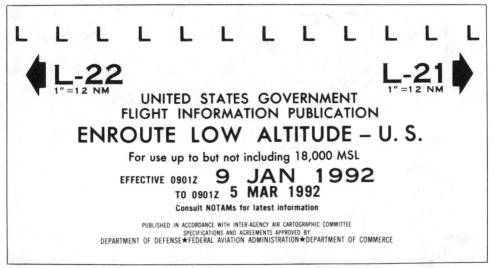

Fig. 12-2. *Another aid for VFR navigation is the Enroute Low Altitude Chart (ELAC).*

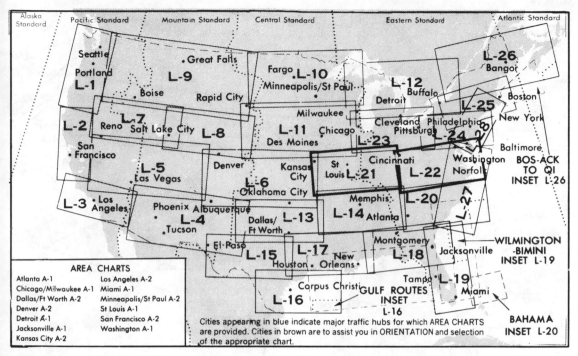

Fig. 12-3. *The 28 charts that cover the contiguous 48 states.*

Twenty-eight charts cover the country, each valid for a period of about two months. These are not 28 individual charts but rather two in one, as FIG. 12-3 indicates. L-21 extends from western Missouri east to portions of Kentucky. L-22 overlaps L-21 in Kentucky and continues east over parts of West Virginia and Virginia to the Atlantic coast.

It's called "Enroute Low Altitude" because it is for use up to but not including 18,000 feet msl. For the VFR pilot who wants to track VORs and use his radio effectively, it provides a wealth of information not found on the sectional. Figure 12-4 illustrates some of the typical data. Keep in mind that it is an enroute chart and thus omits some information found on the sectional, such as unicom frequencies. Nor does it depict topography, towns, roads, rivers, and similar landmarks.

Despite the absence of that information, the chart can almost totally replace the sectional during the enroute portions of a flight if you want to rely primarily on your navcom equipment for navigation. Indeed, it is for IFR flight and contains a lot of symbols that aren't pertinent to the VFR pilot. That fact doesn't reduce its usefulness, however.

I'm not suggesting that you discard the sectional, only that you use the two jointly. After all, you might lose radio contact and have to rely on the sectional to get you to

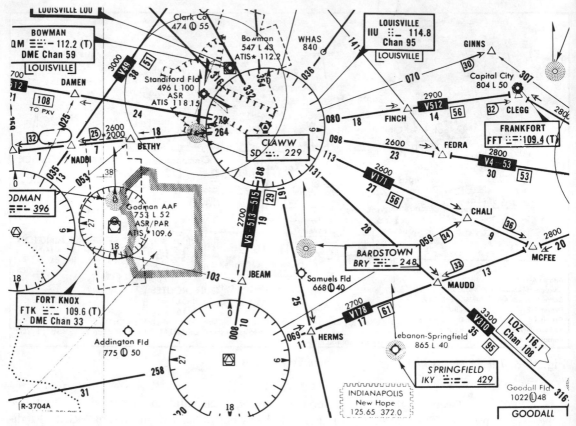

Fig. 12-4. *A sample of what the ELAC shows—and doesn't show.*

the nearest airport. It's a good idea to use the charts in combination so that your position relative to identifiable ground references is never in question.

What do you find in the enroute that's not included in the sectional? A few examples are illustrated in FIGS. 12-5 through 12-9. These represent some of the more meaningful data for the VFR pilot that make the chart a helpful reference and navigation resource. It's not an essential tool, and if you're going to travel strictly by dead reckoning, leave it home. Otherwise, plan your flight with both charts, and use both as you venture from here to there.

THE ADVANTAGES OF CENTER FOR THE VFR PILOT

First, it should be made clear that enroute assistance from Center is neither mandatory nor always available for the VFR pilot. The primary purpose of the ARTCC is to facilitate the movement of IFR aircraft. It does not exist nor is it required to serve the VFR pilot. You can request enroute VFR advisories, but the controller has the

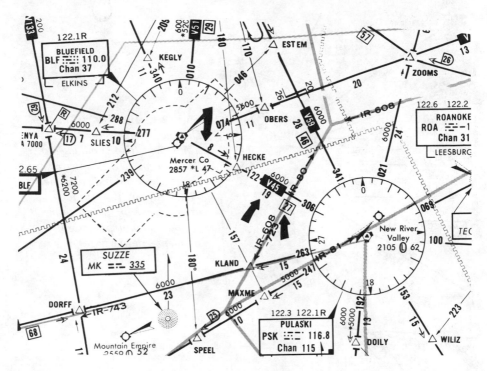

Fig. 12-5. *A helpful feature of the ELAC for the VFR pilot is the listing of nautical miles between VORs, airway intersections, and other IFR reporting points. The "8" is the miles between the Bluefield VOR and the Hecke reporting point; "19" is the distance from Hecke to the Pulaski VOR; the boxed "27" is the total mileage between the two VORs.*

right to refuse the request if his workload does not permit. It is very likely that in marginal VFR conditions or when there is a heavy concentration of traffic, your request for advisories will be rejected. At other times, you will often find Center's controllers not only helpful but eager to be of help.

A second observation: A Center controller is a professional who serves professionals—meaning airline, corporate, and skilled instrument pilots. The vast majority of these airmen know (or should know) how to use their radios. The air-to-ground communications are brief and to the point.

When a VFR pilot gets on the air and proceeds to stammer, hesitate, ramble, or gives the impression of incompetence, the likely response from Center to the request for enroute advisories will be "unable due to workload." That might not be the case at all, but the controller has concluded that he just doesn't have the time to fiddle around with an unprepared amateur. Nor can he be blamed. Remember the case of an airline pilot who reported his position in only five seconds while a private pilot took four minutes to convey the same basic information. Center has neither the time nor patience to accommodate that sort of amateurism.

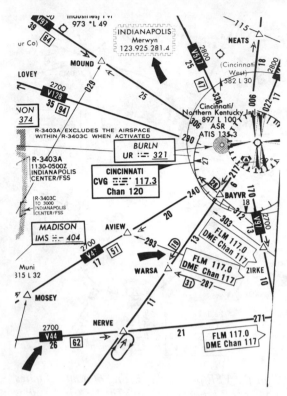

Fig. 12-6. *The "19" is the Distance Measuring Equipment (DME) mileage to Warsa from VOR; the arrow in the lower left indicates the Victor Airway (V44); the arrow at the top points to the remoted Indianapolis Center frequency at Merwyn.*

Fig. 12-7. *The VOR frequency changeover points: 22 miles to the Lawrence VOR and 32 miles to South Boston.*

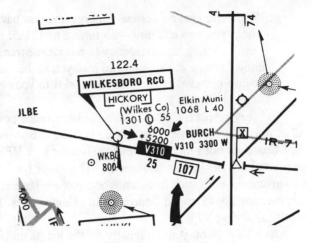

Fig. 12-8. *The minimum enroute altitude (MEA) for radio reception is 6,000 feet msl, and the minimum obstruction clearance altitude (MOCA) is 5,200 feet. The total mileage between IFR reporting points or radio aids (VORs) is "107."*

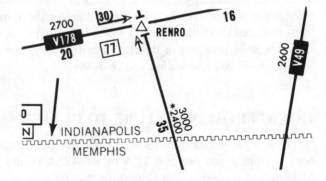

Fig. 12-9. *The boundary line between two Air Route Traffic Control Center areas—in this case, Indianapolis and Memphis.*

None of this is intended to imply that the VFR pilot should avoid using Center or to intimidate the relatively inexperienced. The 100-hour pilot can sound just as professional as the 10,000-hour 747 captain. All it requires is practiced competence in the art of radio communications.

The advantages of using Center on a VFR cross-country are enough to justify the development of professionalism. Let's cite a few:

- The controller with whom you are in contact is, in effect, another pair of eyes—groundbound though they may be. He has your airplane on his scope, identified by the discrete transponder code he has assigned you; thus he can alert you to traffic in your general vicinity. He can issue you warnings of military flight activity, such as B-52s on low-level training flights. If you want to change altitudes, he can advise you of potential traffic at the new altitude you have chosen.

- If you lose all of your radio and squawk the 7700/7600 combination, your Center controller will spot the radio failure, can follow your course and notify subsequent controllers of the situation.

- If you encounter a serious emergency, you have someone on the frequency to whom you can talk now—no tuning to 121.5; no need to divert your attention from handling the emergency to recode the transponder to 7700. The controller knows where you are and can lead you to the nearest airport, keep track of your position, and alert the sources closest to your position of your predicament.

In every respect, Center is an added insurance policy to your flight plan. As with all of the other radio aids, the service is there to be used. I must stress again, however, that Center is not required to lend assistance to a VFR pilot. It's the one case where requested advisories can be rejected. The tower has no such freedom; Approach and Departure have no such freedom when you are landing or departing a TCA or ARSA airport; nor does Ground Control. Only Center does. It must assist the IFR pilot, but not those flying VFR.

As a VFR pilot, you can legally fly the length and breadth of the United States and never once contact any Center. But the proverbial question: Why not use the service, since it's available? At the very least, monitor the remoted frequencies as you pass from one area to another. Merely eavesdropping might alert you to a potential hazard somewhere in your line of flight. It's just as easy, however, to go whole-hog—to go first class—if you know what you're doing.

GOING FROM DEPARTURE TO CENTER

One of three situations relative to Center can occur when Departure Control has been vectoring you out of a TCA or an ARSA (Class B or Class C airspace). The first—rather rare—is when Departure, as you're leaving its area, asks if you would like a "handoff" to Center for enroute traffic advisories. The second is when you initiate the request, Departure asks Center, and Center agrees to provide the service. In the third instance, you again request the handoff, but Center, for whatever reason, denies it. In each scenario, the dialogue follows these general lines:

Situation 1

You are leaving the Kansas City TCA (Class B) VFR for Denver, and about 30 miles out, Departure comes on the air:

Dep: *Cherokee One Four Six One Tango, you are leaving Departure's radar area. Do you want enroute advisories?*

You: Affirmative, Departure, if Center can handle. Six One Tango.

Dep: *Roger, Six One Tango. Stand by.* [Pause, while Departure contacts by telephone the Center sector controller responsible for your geographic area.]

Dep: *Cherokee Six One Tango, contact Kansas City Center, on one two zero point five. Good day.*

You: Roger, one two zero point five. Thank you. Cherokee Six One Tango.

Departure's last communiqué tells you that Center has accepted the handoff and will provide the traffic advisories you have requested. Also implied, but not stated, is the fact that once Center accepted the handoff, Departure gave the sector controller your aircraft type, N-number, present position, altitude, first point of landing, and squawk. In other words, Center knows a lot about you before you make the first contact—which is why that contact is merely "with you," plus your present altitude—or, if climbing, the desired cruising altitude. So, all you do now is leave the discrete transponder code where it is, change to the new frequency, and make the call:

You: Kansas City Center, Cherokee One Four Six One Tango is with you, level at six thousand five hundred.

Ctr: *Roger, Cherokee One Four, Six One Tango. Squawk two two four five and ident.*

You: Two two four five. Cherokee Six One Tango.

Ctr: *Cherokee One Four Six One Tango, radar contact 38 miles west of International. Topeka altimeter three zero one zero. Current traffic C-130s in Topeka area practicing instrument approaches Forbes Field.*

You: Roger, Center, will be looking. Cherokee Six One Tango.

What else follows depends on new traffic and advisories Center might want to communicate to you. Regarding the squawk code, a Center usually, but not always, assigns you a new computer-generated code. Until it does, though, leave the one that Departure gave you right where it is.

Situation 2

Here, Departure terminates radar coverage and makes no reference to Center, but you do want flight-following advisories:

Dep: *Cherokee One Four Six One Tango, radar service terminated. Squawk one two zero zero. Frequency change approved. Good day.*

You: Departure, Cherokee One Four Six One Tango requests handoff to Center, if possible.

Dep: *Stand by, Six One Tango. We'll check . . .* [pause] *. . .*

Dep: *Cherokee Six One Tango, contact Kansas City Center on one two zero point five.*

You: Roger, one two zero point five. Thank you. Cherokee Six One Tango.

That's all there is to it. Merely change the frequency and contact Center, just as in the previous example.

Situation 3

In this instance, you want enroute advisories, Departure requests it, but Center is unable to provide it:

Dep: *Cherokee One Four Six One Tango, radar service terminated, Squawk one two zero zero. Frequency change approved. Good day.*

You: Departure, Cherokee Six One Tango requests handoff to Center.

Dep: *Cherokee Six One Tango, stand by. We'll check . . . [pause] . . .*

Dep: *Cherokee Six One Tango, Center unable at this time. Radar service terminated, Squawk one two zero zero. Frequency change approved.*

You: Roger, squawking VFR. Thanks for your help. Cherokee Six One Tango.

With this, you're on your own as you head west. You can change altitudes as you wish, as long as they conform to the VFR altitude regulations; you can alter your route of flight; you can change your radio to any frequency you wish; and you must remain VFR at all times. In other words, you have the freedom to do just about as you wish—as long as you conform to the VFR FARs.

Any time that you hear the word "terminated," you can assume that all radar service has ceased and that no handoff to the next agency has taken place. Now if you want Center or Approach, you have to give the full IPAI/DS. The next facility hasn't heard about you before.

Once in a while, a controller might misuse "terminated" as it's supposed to be applied per the book. You hear something like this from Departure: "Cherokee Six One Tango, radar service terminated. Contact Atlanta Center on one two four point seven." Has radar service literally been terminated, or have you been handed off to Center? The first part says terminated; the second part raises a question. Uncertain of what has really happened, you play the pro and call Center:

You: Atlanta Center, Cherokee One Four Six One Tango.

Ctr: *Cherokee One Four Six One Tango, Atlanta Center. Go ahead.*

The phraseology indicates that this is the initial contact with the controller, so you go ahead with IPAI/DS.

But suppose you hear this in response to that call:

Ctr: *Cherokee One Four Six One Tango, Atlanta Center. Squawk one two zero five and ident. Verify altitude.*

You: Cherokee One Four Six One Tango, squawking one two zero five. Level at seven thousand five hundred.

This time the phraseology tells you that you have been handed off. In either case, once contact with Center has been established, the controller takes care of much of the

conversation from that point on. Your basic job is to listen, respond to
edge whatever information or instructions are conveyed.

Four Points about Handoffs—in Review

- The expression "with you" is used any time you are being handed off or automatically transferred from one agency to another: Departure to Center; Center
  to Approach; Approach to Tower.

- Even though it is a handoff, be sure to advise the receiving agency of your
  present altitude. If you fail to include it in your initial contact, you might be
  asked—which only adds to air clutter.

- Note that no position report is necessary. Center, or whatever the controlling
  agency is, knows your position as well as your destination and squawk.

- Don't change your transponder from one squawk to another until you are so
  advised. Let's say that Departure has given you 1201 and you have been handed
  off to Center. Keep 1201 in the box unless Center gives you a new code.

INITIATING THE CONTACT WITH CENTER

Either Departure hasn't handed you off or you're leaving an airport that doesn't
have a Departure Control. Regardless of the situation, you want to establish the first
contact with a Center for enroute advisories. You've determined the probable Center
frequency through proper preflight preparation, so, again, it's the simple IPAI/DS:

You: Seattle Center, Cherokee One Four Six Tango.

Ctr: *Cherokee One Four Six One Tango, Seattle Center.*

You: Center, Cherokee Six One Tango is about 15 south of Astoria on Victor Two
Seven, level at five thousand five hundred, VFR to Newport. Request traffic
advisories, workload permitting.

Ctr: *Cherokee One Four Six One Tango, squawk four one four five and ident.*

You: Four one four five. Cherokee Six One Tango.

Ctr: *Cherokee Six One Tango, radar contact eighteen south of Astoria. Maintain VFR at all times. Astoria altimeter two niner seven five. Seattle
Center.*

You: Roger, two niner seven five. And will maintain VFR. Cherokee Six One
Tango.

That's all there is to it. Center will take it from there and keep you advised of
what's going on around you.

About that comment, "workload permitting." It's obviously not required, any
more than "good day" or "thank you," but it does put your request in the form of a
request, not a command. In its way, it indicates appreciation of the fact that the controller might be busy and not able to provide the advisories. That little added phrase

frequently is enough to get the help that otherwise might have been rejected. Call it courtesy and mutual understanding in the air.

Also, be sure that you have established contact with the Center before giving the IPAI/DS, as the initial transmission above indicates. The controller might be busy on another frequency and won't hear your call, so don't waste air time until you know you've got a listener.

Now that Center has you on radar and is tracking your progress, an admonition is in order: Even with Mode C, don't change altitudes without advising Center. The controller has you pegged at 5,500 feet, and if you don't have Mode C, there's no way he can determine your altitude without verification from you. Even with Mode C, an unapproved change to a different flight level might affect the flow of traffic and place you and others in jeopardy. You're not flying under instrument flight rules, and you do have the VFR freedom to vary your altitude. But you've told Center one thing, so don't make changes without communicating your intended actions.

ENROUTE FREQUENCY CHANGES

You're moving along over the countryside, let's say, from Kansas City to Memphis. You've already made the initial contact with Center, as in the previous Seattle example, when you hear something like this:

Ctr: *Cherokee One Four Six One Tango, contact Kansas City Center, frequency one two five point three.*

You: Cherokee Six One Tango, Roger, one two five point three.

The call simply means that you're leaving the controller's sector. So you "good day" him and tune in 125.3. Then make contact on the new frequency:

You: Kansas City Center, Cherokee One Four Six One Tango with you, level at seven thousand five hundred.

All you're doing is talking to a different controller at Kansas City Center who is handling the geographical area you are now in.

Along the same line, you might hear this:

Ctr: *Cherokee Six One Tango, change to my frequency* [or "contact me now on"] *one one eight point five five.*

You: Roger. One one eight point five five. Cherokee Six One Tango.

Change your radio and make contact again: "Kansas City Center, Cherokee Six One Tango is with you on one one eight point five five."

Notice the difference between the calls? In the latter case, the controller said ". . . change to my frequency . . ." It's the same person you've been chatting with all along. He just wants you on a different remote site frequency. There's no need for the traditional "good day" or altitude report. You haven't left him. Listen for the inclusion of "me" or "my" in the controller's message. That's the tipoff.

ENROUTE ADVISORIES AND CENTER

Back to the flight itself: You're with Center and squawking the code given you. What the controller passes on to you now depends on his workload and the conditions along your route of flight. Typical of what you might hear are the following examples:

Example 1

Ctr: *Cherokee Six One Tango, traffic at ten o'clock, three miles, northeast bound. Altitude unknown.*

You: Negative contact. Cherokee Six One Tango. [Or, Cherokee Six One Tango has the traffic. Or, Cherokee Six One Tango requests vectors around the traffic.]

Some old-time pilots still use the phrase "no joy" instead of "negative contact," and "tallyho" instead of "has the traffic." Neither term, however, is part of the approved phaseology, and is thus frowned on by the FAA.

Example 2

Ctr: *Cherokee Six One Tango, were you advised of low-level military flights in your present area?*

You: Negative, not so advised.

Ctr: *Cherokee Six One Tango, be alert for north-south B-52 activity.*

You: Cherokee Six One Tango will be looking. Thank you.

Example 3

You see ahead of you a cloud buildup that appears to be right at your altitude. To avoid it, you decide to drop down from 7,500 to 5,500 feet. That's permissible, but advise Center ahead of time of your plans:

You: Center, Cherokee Six One Tango descending to five thousand five hundred due to clouds.

Ctr: *Cherokee Six One Tango, Roger. Report reaching five thousand five hundred.*

You: Cherokee Six One Tango wilco.

You: Center, Cherokee Six One Tango level at five thousand five hundred.

Ctr: *Cherokee Six One Tango, Roger.*

Example 4

You're nearing Springfield, which, although it has no TCA or ARSA, does have Approach/Departure Control. A few miles out, Center comes on the air:

Ctr: *Cherokee One Four Six One Tango, contact Springfield Approach on one two four point niner five. Good day.*

You: Roger, Center, one two four point niner five. Thanks for your help. Cherokee Six One Tango.

A logical question: You're not landing at Springfield on this flight to Memphis, so why call Approach? The answer: Simply because you're about to enter Approach's area of radar coverage—its area of responsibility—which has a horizontal radius of about 30 miles and rises vertically to 10,000 or 12,000 feet. Once you're through the area, Approach will either hand you off to Center again (which is probable) or advise you that radar service is terminated.

Example 5

You are leaving Kansas City Center's jurisdiction and approaching that of the Memphis Center:

Ctr: *Cherokee Six One Tango, contact Memphis Center now one two four point three five.*

You: Roger. One two four point three five. Cherokee Six One Tango. Good day.

Note that Kansas City has said nothing about radar service being terminated. This means that you have been handed off to Memphis. Just change to 124.35 and make your call:

You: Memphis Center, Cherokee One Four Six One Tango is with you, level at five thousand five hundred.

Ctr: *Roger. Cherokee Six One Tango. Altimeter three zero one five. Continue present heading* [or whatever the instructions, if any, might be].

A comment or two about frequency changing is appropriate here. Let's say that you have two navcoms. The last Kansas City frequency, 125.3, is in the first navcom. When told to change to Memphis, enter 124.35 in the second navcom, but leave the first where it is. If for any reason you have to call Kansas City again, you can do so without time-consuming dial changes. Once contact is made with Memphis, merely set the first navcom to the next probable Memphis remote frequency, or turn it to any other radio aid you want.

With only one navcom aboard, be sure to write down the new frequency given you. (It's easy to forget or confuse 124.35 with 123.45—or any other combination.) If you've done this progressively from the first Departure frequency through the one or more Center frequencies, the piece of paper on your knee pad will resemble this:

KC D/C	119.0
KC Ctr	125.55
KC Ctr	125.3
MEM Ctr	124.35

Just cross out each previous frequency, but don't obliterate it. You might need it again.

Even with two navcoms, the practice of writing down each succeeding frequency makes good sense. It's a little embarrassing, after a minute or two of uncertainty or forgetfulness, to have to recontact the last controller and ask: "What frequency did you tell me to change to?"

Example 6

As the flight progresses, you find that the headwinds are much stronger than forecast. Your forward progress is slower than the rate at which the fuel gauges are dropping, so you decide to put down at Springfield, Missouri. Because you're still in Kansas City Center's area and they have you on a nonstop flight to Memphis, a call to Center is essential:

You: Center, Cherokee Six One Tango will be landing Springfield for fuel.

Ctr: *Cherokee Six One Tango, Roger. Contact Springfield Approach on one two four point niner five.*

You: Roger. One two four point niner five. Cherokee Six One Tango. Good day.

Assuming that you filed a flight plan before leaving Kansas City, it's almost certain that the ground time at Springfield will delay your original ETA at Memphis. Accordingly, be sure to telephone the Flight Service Station and either file a new flight plan or ask that your ETA be revised. Otherwise, the FSS will start asking questions 30 minutes after that first ETA expires.

CENTER TO APPROACH

Airborne again (or maybe you didn't have to stop at all), you're nearing Memphis International. At a given point, Memphis Center will contact you with one of three possible instructions.

Example 1

Ctr: *Cherokee Six One Tango, radar service terminated. Contact Memphis Approach on one two five point eight.*

You: Roger, one two five point eight. Cherokee Six One Tango. Good day.

Center said "terminated," so you can assume that you are not being handed off. But the controller gave no instructions about changing the transponder to 1200 or any other code, which means that you leave it at the last setting—1205, or whatever it was. Then call Approach with the standard IPAI/DS:

You: Memphis Approach, Cherokee One Four Six One Tango.

App: *Cherokee One Four Six One Tango, Memphis Approach.*

You: Cherokee One Four Six One Tango is with you over Marion, level at five thousand five hundred, landing Memphis. Squawking one two zero five with Information Quebec.

Approach will take it from there.

Example 2

Ctr: *Cherokee Six One Tango, radar service terminated. Squawk one two zero zero. Contact Memphis Approach on one two five point eight.*

You: Roger, one two five point eight. Cherokee Six One Tango. Good day.

With this advice, you know there has been no handoff, so change to 1200, and give Approach the full IPAI/DS.

Example 3

Ctr: *Cherokee Six One Tango, contact Memphis Approach on one two five point eight.*

You: One two five point eight. Cherokee Six One Tango.

You: Memphis Approach, Cherokee One Four Six One Tango is with you, level at five thousand five hundred.

This was a direct handoff. Approach knew all about you and your intentions—thus no need for IPAI/DS, other than to confirm your present altitude.

Using Center as Approach/Departure Control

As mentioned in chapter 11, Center offers limited Approach and Departure Control for many airports that would otherwise have no approach/departure services, including:

- Nontower airports that are not within radar coverage of a larger airport's Approach/Departure Control

- Tower airports where activity levels do not justify airport-based radar and which are not within the radar coverage of a larger airport's Approach/Departure Control

- Airports with a part-time tower and part-time Approach/Departure Control when those services are closed

- Tower and nontower airports within radar coverage of a larger airport's part-time Approach/Departure Control when such service is closed.

Like the other Center services discussed, VFR Approach and Departure services are available from Center on a workload-permitting basis only. Figures 12-10 through 12-12 are just three examples of airports with Center-supplied Approach and Departure services.

```
CONCORD MUNI    (CON)   2 E   UTC–5(–4DT)   43°12'12"N 71°30'09"W              NEW YORK
   346   B   S4   FUEL 100LL, JET A                                           H-3J, L-25D, 26F
   RWY 17-35: H6005X100 (ASPH)   S-43, D-60   HIRL                                  IAP
      RWY 17: REIL. PAPI(P4L). Thld dsplcd 640'. Trees.    RWY 35: MALSR. VASI(V4L)—GA 3.0°TCH 51'.
   RWY 03-21: H3999X150 (ASPH)   S-30
      RWY 03: Thld dsplcd 440'. Trees.        RWY 21: Trees.
   RWY 12-30: H3499X150 (ASPH)
      RWY 12: Trees.        RWY 30: Trees.
   AIRPORT REMARKS: Attended Nov-Apr 1200-2200Z‡, May-Oct 1230–0000Z‡. Rwy 03–21 CLOSED winter months.
      ACTIVATE HIRL Rwy 17–35 and MALSR Rwy 35—CTAF. Rwy 17 REIL out of svc indefinitely.
   COMMUNICATIONS: CTAF/UNICOM 122.7
      BANGOR FSS (BGR) TF 1–800–WX–BRIEF. NOTAM FILE CON.
      RCO 122.3 122.2 (BANGOR FSS)
      ®MANCHESTER APP/DEP CON 127.35 (1100-0500Z‡)   CLNC DEL 133.65
      ®BOSTON CENTER APP/DEP CON 123.75 (0500-1100Z‡)
   RADIO AIDS TO NAVIGATION: NOTAM FILE CON. VHF/DF ctc BANGOR FSS
      (L) VORTACW 112.9   CON   Chan 76   43°13'11"N 71°34'33"W   122° 3.4 NM to fld. 710/15W. HIWAS
      EPSOM NDB (MHW/LOM) 216   CO   43°07'07"N 71°27'11"W   353° 5.5 NM to fld. Unusable beyond 20 NM.
      ILS 108.7 I-CON Rwy 35. LOM EPSOM NDB.
```

Fig. 12-10. *Concord, New Hampshire, a nontower airport, receives Approach/Departure Control service from Manchester, New Hampshire, or from Boston Center when Manchester Approach is closed.*

```
MANCHESTER    (MHT)   3 S   UTC–5(–4DT)   42°56'00"N 71°26'18"W              NEW YORK
   234   B   S4   FUEL 100LL, JET A   OX 3   LRA   ARFF Index C              H-3J, L-25D, 26F
   RWY 17-35: H7001X150 (ASPH-GRVD)   S-200, D-200, DT-350   HIRL                   IAP
      RWY 17: REIL. VASI(V4L).        RWY 35: MALSR.
   RWY 06-24: H5847X150 (ASPH)   S-200, D-300, DT-350   HIRL
      RWY 06: VASI(V4L).        RWY 24: REIL. VASI(V4L).

   WEATHER DATA SOURCES : LAWRS.
   COMMUNICATIONS: CTAF 121.3   ATIS 119.55 (1100–0500Z‡)   UNICOM 122.95
      BANGOR FSS (BGR) TF 1–800–WX–BRIEF. NOTAM FILE MHT.
      RCO 122.1R 114.4T (BANGOR FSS)
      ®APP/DEP CON 124.9 134.75 (1100-0500Z‡)    ® BOSTON CENTER APP/DEP CON 123.75 (0500-1100Z‡)
      TOWER 121.3 (1100-0500Z‡)   GND CON 121.9    CLNC DEL 135.9
      ARSA ctc APP CON 127.35 (North) 124.9 (South)
   RADIO AIDS TO NAVIGATION: NOTAM FILE MHT.
      (L) VORTAC 114.4   MHT   Chan 91   42°52'06"N 71°22'12"W   337° 4.9 NM to fld. 470/15W
      DERRY NDB (MHW) 338   DRY   42°52'12"N 71°23'52"W   351° 4.2 NM to fld.
      ILS 109.1 I-MHT Rwy 35
```

Fig. 12-11. *Manchester has its own Approach/Departure Control, but when closed, Boston Center provides the service.*

CHARLOTTESVILLE-ALBEMARLE   (CHO)   7 N   UTC–5(–4DT)   38°08'19"N 78°27'11"W   WASHINGTON
  641   B   S4   FUEL 100LL, JET A   ARFF Index B   H-6H, L-22G, 27D
  RWY 03-21: H6001X150 (ASPH-GRVD)   S-100, D-160, DT-300   HIRL   0.3% up SW   IAP
    RWY 03: MALSR (unmonitored). Tree.   RWY 21: REIL. VASI(V4L)—GA 3.0°TCH 50'.
  AIRPORT REMARKS: Attended 1100-0400Z‡. PPR 8 hours for unscheduled air carrier ops with more than 30
    passenger seats call arpt manager 804–373–8341. When twr clsd ACTIVATE MALSR Rwy 03 and HIRL Rwy
    03–21—CTAF. Deer/dogs/birds on and in vicinity of arpt. Men and equipment working adjacent to Rwy 03–21,
    north and south parallel taxiways indefinitely. VASI Rwy 21 not monitored. REIL not monitored. Landing fee only
    for acft over 7000 lbs. Wheeled helicopters are requested to ground taxi when transitting arpt. Control Zone
    effective 1148-0400Z‡.
  WEATHER DATA SOURCE: AWOS–3 118.425 (804) 973–4403. (Ops hours twr clsd)
  COMMUNICATIONS: CTAF 124.5   ATIS 118.425 (Opr twr hours)   UNICOM 122.95
    LEESBURG FSS (DCA) TF 1–800–WX–BRIEF. NOTAM FILE CHO.
    CHARLOTTESVILLE RCO 122.65 122.2 122.1R (LEESBURG FSS)
➤ ®WASHINGTON CENTER APP/DEP CON 124.25
    TOWER 124.5 (1148-0400Z‡)   GND CON 121.9
  RADIO AIDS TO NAVIGATION: NOTAM FILE DCA.
    GORDONSVILLE (H) VORTAC 115.6   GVE   Chan 103   38°00'48"N 78°09'12"W   304° 16.1 NM to fld. 380/06W.
    HIWAS.
    AZALEA PARK NDB (MHW) 336   AZS   38°00'36"N 78°31'06"W   030° 7.8 NM to fld. NOTAM FILE CHO.
    NDB unmonitored when twr clsd.
    ILS 111.7 I-CHO Rwy 03 LOC/GS unmonitored when twr closed. MM/OM unmonitored.

Fig. 12-12. *At Charlottesville, Virginia, Washington Center is responsible for Approach/ Departure—even when the Charlottesville Tower is open.*

CONCLUSION

That's pretty much the story of Center, as far as the VFR pilot is concerned. To summarize some of the points I've made:

- Center exists primarily to serve IFR flights.

- Center's assistance to VFR pilots is on a workload-permitting basis.

- Use of Center by the VFR pilot is advisable but not mandatory.

- Although the VFR pilot is not under Center's control, he should not deviate from announced altitudes or routes of flight without advising Center in advance.

- Become familiar with the probable frequencies you will use via the ELAC and know approximately when you will be asked to change from one remoted frequency to another or from one Center to another.

- Become familiar with the ELAC for more exact radio navigation and frequency-change areas.

- Write down all frequencies (planned or given) in chronological sequence so you won't forget or become confused.

- Plan your communication to Center before picking up your mike.

- Rehearse your message so that you sound like a pro—not a hesitant amateur.

- Center is one more eye in the sky to help you get where you want to go. I urge that you use this control source, when possible, for that added insurance. I urge that you use it on every cross-country flight. Too many non-IFR pilots, because of lack of knowledge and confidence, are afraid of it—when there is no reason for fear of any sort. If the controller can't accommodate you, he'll tell you; if you don't understand a direction, ask him to repeat or clarify. I've heard airline pilots, who are presumably pros, do this on any number of occasions. In every case, the controller honored the request without rancor or intimidation. He'll do the same for you if you sound as though you know what you're doing and what's going on.

One final observation: there's an old saying that "what we're not up on, we're down on." What we don't understand, we're either against or we fear. Uncertainty breeds insecurity. This chapter (and, in fact, the whole book) tries to provide some knowledge that will lead to understanding and greater pilot confidence in the field of radio communications.

Rarely can examples and the written word accomplish the entire task, however. Consequently, I recommend that you visit the Center nearest you. Call a supervisor, explain what you want, and request a brief tour of the facility. See what's going on. Talk to a couple of controllers. Listen to the communications between ground and air. Study the screens and the blips that identify the various aircraft.

Firsthand experience is the best way I know to bring meaning to words and examples to life. The FAA urges all pilots to visit its facilities—Center, Tower, Flight Service, and the rest. The facilities are glad to welcome you, and, workload permitting, will see that your tour is complete and educational. It's worth a couple of hours of your time to put to rest any feelings of uncertainty. "What we're up on, we're not down on." The saying works in reverse as well.

13

In the Event of
Radio Failure

Despite the sophistication of modern avionics, things can go wrong. When your radio decides to take a rest, it is comforting to know what you need to do to get your airborne vehicle safely back on the ground.

As *AIM* says, it's virtually impossible to establish fixed procedures for all situations involving two-way radio failure. Preferred actions, however, can be outlined. Within those parameters, every pilot should have a plan in the unlikely event communications are lost. Just as the intelligent pilot is mentally prepared for an engine failure, so is he ready to cope with a radio failure.

ONE PREVENTIVE STEP

It's pretty hard to determine the potential for radio failure during the preflight check. Of course, a loose or broken antenna is an obvious signal, as is a navcom that slips in and out of its housing rack, a circuit breaker that has popped, or a frayed magneto wire. Otherwise, there's not a lot you can check.

One symptom of a potential problem, however, is the alternator belt. If you can inspect it visually, check it for wear and frays. Whether you can see it or not, check its tension. If it's loose, have it replaced or tightened. Not all radio failures can be traced to alternators or alternator belts, but the malfunctioning of one or the other will result in a general power loss, with the battery supplying what little remaining electrical juice it can generate. And the battery won't last forever under those conditions.

WHEN A FAILURE IS SUSPECTED

Every once in a while, the radio might give off an eerie silence; what has been a pattern of constant chatter suddenly is no more; or your calls to a ground facility seem to fall on deaf ears; whatever the case, you begin to wonder if . . . .

Before getting too concerned, adjust the squelch or turn up the volume on the radio. If you get the characteristic static, the radio is working. It's just that there's been an unusual dearth of communication activity over the past few minutes. If there is no static sound, push the set in a little. Vibration might have caused it to slip out of its rack just enough to break electrical contact. Should you appear not to be transmitting, check all connections. Is the mike plugged in? If you are using a headset, is it plugged in? Little things can happen in normal operations that produce the symptoms of radio failure, but they can be corrected with only a minor adjustment.

However, when a previously clear reception starts to break up, or you develop an unusual hum in the speaker, trouble might be brewing. Check the ammeter. Is it still showing a charge? Check the circuit breakers. If a navcom button has popped, let it cool off for a couple of minutes before resetting it. Does that do the trick? If so, you're probably all right . . . for a while. However, something is shorting out and should be checked as soon as you're back on the ground.

But suppose the ammeter shows no charge at all. Test it again by turning on the landing light. If the needle doesn't move, you can be sure the alternator has died or its belt has broken. In either case, what electrical power remains is coming from the battery alone. This being the case, turn off *all* nonessential electrical equipment, except one radio (assuming you can get some reception over it), and head for home or the nearest airport.

Unless you have experienced a radio or electrical failure in flight, the ammeter is probably one of your least monitored instruments. I suggest that it be permanently incorporated in the panel scanning process. The sooner a problem is caught, the better, because the life of a battery, once the alternator is gone, is about two hours. After that, you'll have no electrical power for lights, navcoms, transponder, gauges, and the like. The engine won't stop, because the ignition system is independent of the alternator-battery system—but a functioning engine, as critical as that is, is about all you'll have left.

WHEN A FAILURE IS CONFIRMED

There is no question: The failure is real. The reception is getting weaker and weaker and your transmission is almost unreadable. To explain your options, let's set up five VFR situations—four in the air and one on the ground—that you might someday encounter.

Case 1:

This example is the simplest of all. You're flying locally out of an uncontrolled unicom or multicom airport or you're on a short cross-country with no intermediate landing planned. Because two-way communications are not required under these conditions, there's no need to alert ATC with the standard radio failure transponder codes when the set goes bad. Just continue to squawk 1200, stay away from TCAs, ARSAs, and ATAs, and land at your own or some other uncontrolled field to get the radio fixed. Do listen, though, for CTAF traffic advisories, if the receiver works at all. And do be careful as you enter the traffic pattern. You're coming in unannounced and you might not know who else is in the pattern or where he is—both of which are conditions that open the door to trouble.

Case 2:

This time your destination is the primary airport in a TCA (Class B). The symptoms of a sick radio have been plaguing you for the past hour or so, but there's enough juice left to monitor the ATIS, contact Approach Control, announce your intentions, and get clearance into the TCA. Once inside, however, the radio becomes useless. You can't decipher what Approach is communicating and Approach can't understand your transmissions. Now what do you do?

First, as soon as you're aware of the failure, squawk 7700 for one minute. (Be sure you time it.) The flashing "7700" on the radar screen immediately alerts Approach, the nearest Center, and a radar—equipped tower that you've got a problem. After one minute, change the squawk to 7600 and continue that transmission for 15 minutes. When the "7600" code appears on the screens, the controllers know that the problem is radio failure (RF)—not a distress or urgency situation. Then, if you're still in the air after 15 minutes, repeat the 7700/7600 sequence. (This repetition applies any time you have a radio failure, regardless of where you are.)

Second, continue through the TCA to the airport environment, Approach knows what you intend to do and will protect you just as it would an IFR aircraft.

Third, knowing from the ATIS the active runway, winds, and so on, enter the ATA with the landing pattern in mind. As you do, however (and this is most important), watch the tower for light gun signals that will clear you to land or tell you to keep circling. (These signals are discussed further in Case 3 and are illustrated in TABLE 13-1.) When you see the green light, that's your landing clearance. Once on the ground and off the active runway, continue to monitor the tower for the flashing green light that clears you to taxi.

Now get the radio fixed. The odds are almost certain that you won't be allowed to depart the TCA airport until two-way communications are reestablished.

The above scenario applies equally:

- To radio failure in an ARSA
- If your destination is a satellite airport that underlies but is not in a TCA or ARSA and you are already in either of the airspaces
- If you're simply transiting a TCA or an ARSA.

In essence, once you've made contact with Approach and have been cleared into

Table 13-1. Tower light signals

COLOR AND TYPE OF SIGNAL	MEANING		
	MOVEMENT OF VEHICLES EQUIPMENT AND PERSONNEL	AIRCRAFT ON THE GROUND	AIRCRAFT IN FLIGHT
Steady green	Cleared to cross, proceed, or go	Cleared for takeoff	Cleared to land
Flashing green	Not applicable	Cleared for taxi	Return for landing (to be followed by steady green at the proper time)
Steady red	STOP	STOP	Give way to other aircraft and continue circling
Flashing red	Clear the taxiway/runway	Taxi clear of the runway in use	Airport unsafe, do not land
Flashing white	Return to starting point on airport	Return to starting point on airport	Not applicable
Alternating red and green	Exercise extreme caution	Exercise extreme caution	Exercise extreme caution

the TCA or ARSA, keep right on going in accordance with your announced intentions. That's much safer than wandering around in a sort of panic because you've lost radio contact with the ground. When you deviate from what you've told the controller, he doesn't know what to expect—which could cause a fair amount of confusion in his efforts to maintain an orderly flow of traffic.

Case 3:

This time you want to land at a tower-controlled airport that has no radar, has no local Approach Control facility, and is miles away from either a TCA or an ARSA. The radio is dead, but you squawk the usual 7700/7600 RF code. Considering the circumstances, of what value is that squawk, and what do you do now?

Well, the nearest Center (or its remoted outlet) or Approach, seeing the radio-failure (RF) code flashing on its screen, will start tracking you on radar. If it then appears that you're heading for the Tower airport, one of the facilities will notify the tower by phone of your probable landing intentions.

As you near the airport, try to establish contact. Just possibly there's enough life in the battery or the radio to get the necessary pattern instructions and landing clearance. If not, cross over the field at least 500 feet above pattern altitude, observe the flow of traffic and the active runway, and be very much on the lookout for other aircraft.

While still well above pattern altitude, fly upwind over the active runway, the purpose being to attract the controller's attention. If Center or Approach has alerted him to the arrival of an RF aircraft, he'll be watching and will try to reach you by radio. Should the radio be working at all, you'll hear something like: "Aircraft over the runway, Albany Tower. If you receive me, rock your wings." This you do, followed by: "Roger, do you intend to land at Albany?" Again rock the wings. The controller will then fit you into the pattern and give you landing clearance.

If the radio is completely dead, you might have to fly over the runway a couple of times, but be sure to keep an eye on the tower. The controller, knowing that he's not in radio communication, will instruct you via his light gun by aiming the gun at your aircraft and flashing one of the signals listed on TABLE 13-1—most probably a steady green or a steady red.

The only problem about seeing the light is the location of the tower in relation to the landing runway. You might have difficulty seeing it if you're in the left seat and the tower is on your right. Should that be the case, watch for the light on the crosswind or downwind leg—still 500 feet or so above pattern altitude. Once you've got the steady green, rock your wings in acknowledgment.

Don't fail, though, to check the tower frequently for any subsequent visual instructions. Something might happen, especially on the final, that would cause the tower to want you to abort your landing—an unauthorized aircraft or a ground vehicle on the runway, you forgot to lower your gear, a sudden accident, a jet closing rapidly behind you, an aircraft cutting in ahead of you, the spacing between aircraft is too close, or whatever. With no radio, the gun is the only means of communication the tower has with you, so be alert to the fact that you might get a steady or flashing red light any time in the pattern.

Case 4:

You want to land at the primary airport in a TCA or ARSA. You have not been using Center and the radio fails before you have contacted Approach. Your only alternative in this situation is to land somewhere outside of the TCA or ARSA and then request entry approval from Approach by telephone. Although highly unlikely without two-way radio communications, Approach might grant permission if the traffic is light and you are within a few miles of the primary field.

Case 5:

You're on the ground at some uncontrolled field, perhaps your home base. The radio is dead and you want to fly to a tower-controlled airport, which either underlies or is far from a TCA or ARSA, for repairs. What are your options?

Really, you have only one: A telephone call to the tower supervisor before departing. In the call, explain your predicament, estimate the time of arrival in the area, describe your aircraft, and request approval to enter the ATA. If the supervisor can grant your request, he'll probably give you instructions about pattern altitudes and what to do when you get into the ATA.

Since you're already on the ground, there's no emergency. Permission to enter the

ATA is thus a matter of option, largely depending on probable traffic volume, weather conditions, field conditions, or other factors that might affect approval or denial of the request. It's reasonable to assume that the tower will do what it can to help you, but that assistance isn't automatic. Two-way radio communications are too critical in a busy terminal environment—which means that you might have to look elsewhere for a repair shop at some uncontrolled field.

CONCLUSION

The preceding recommendations, suggestions, or procedures, are fine; they will *almost* always get you down without any problems. But they don't work 100 percent of the time.

I knew a pilot who once lost an alternator belt in the vicinity of a TCA and an underlying radar-equipped airport. He could receive, although the reception was scratchy and getting worse, but not transmit. Tuned to the tower and squawking 7700/7600 produced no response, so he flew up the landing runway at 2,000 feet agl.

Midway over the airport, he heard: "Aircraft over the runway, rock your wings if you receive me and are landing at [*name of airport*]."

Replying affirmatively, he was sequenced into the pattern and cleared to land.

Once off the runway, he switched to Ground, but if any taxi clearance was transmitted, the radio was too far gone for him to know. And there was no light from the tower.

After a couple of futile minutes, he taxied slowly to the ramp and placed a telephone call to the tower to explain his unauthorized taxiing and to ask if the 7700/7600 transmission had been received. The response was no—because the radar was down. Why no light signals? Sorry about that, but "our gun is broken."

So, not all malfunctions involve the aircraft. In this case, it was the alertness of the controller who spotted him above the runway and led him down to an uneventful landing at a busy airport.

Radio failures are relatively rare, but when they do occur, you should know what to do. The best advice: Don't bust into a controlled airport with no radio unless you feel that you have no safe alternative. Get out of the vicinity, find an uncontrolled field, and have the problem repaired (or make a phone call to the tower to request a no-radio clearance).

14
A Cross-Country: Putting It All Together

The preceding chapters have taken you from multicom through Center. Now let's put it all together with a mythical cross-country trip. Except for multicom airports, I try to incorporate at least one example of the radio communications with each of the various facilities I've been discussing. I hope, then, the "flight" will serve as a reasonable model for the real-life excursions you might make. Naturally, your itineraries will differ, as will the frequencies, but the principles illustrated should not vary because of that.

The flight will take you from the Kansas City Downtown Airport to Omaha's Eppley Field, where you'll pick up two friends. From there, you'll go to the Minneapolis International Airport. After completing your business in the Twin Cities, you'll drop off one of your friends at Mason City, Iowa, then the other friend at Newton, Iowa. From Newton, it's nonstop back to Kansas City.

The entire flight presupposes that you have a Mode C transponder, two navcoms, distance measuring equipment (DME), and no loran. If you don't have the luxury of multiple navcoms, the principles are the same, but frequency-changing is a little less convenient.

THE FLIGHT ROUTE

Without loran, you decide to fly the airways whenever possible, even though doing so will add a few miles to the trip. Accordingly, the route of flight, VORS, course headings, and point-to-point nautical mileage resemble FIG. 14-1.

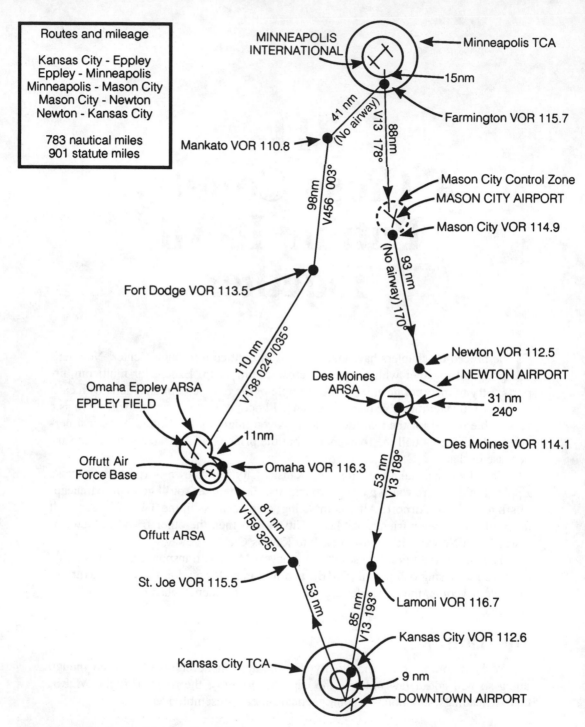

Routes and mileage

Kansas City - Eppley
Eppley - Minneapolis
Minneapolis - Mason City
Mason City - Newton
Newton - Kansas City

783 nautical miles
901 statute miles

MINNEAPOLIS INTERNATIONAL

Minneapolis TCA

15nm

Farmington VOR 115.7

41 nm (No airway)

V13 178° 88nm

Mankato VOR 110.8

Mason City Control Zone
MASON CITY AIRPORT
Mason City VOR 114.9

98nm V456 003°

V13 93 nm (No airway) 170°

Fort Dodge VOR 113.5

Newton VOR 112.5
NEWTON AIRPORT

31 nm 240°

110 nm V138 024°/035°

Des Moines ARSA

Des Moines VOR 114.1

Omaha Eppley ARSA
EPPLEY FIELD

11nm

Offutt Air Force Base

Omaha VOR 116.3

53 nm V13 189°

Offutt ARSA

V159 325° 81 nm

St. Joe VOR 115.5

53 nm

85 nm V13 193°

Lamoni VOR 116.7

Kansas City TCA

Kansas City VOR 112.6

9 nm

DOWNTOWN AIRPORT

Fig. 14-1. *The route of the simulated flight.*

As 900 statute miles is obviously too much territory to cover in one day, with stops and business enroute, you plan to stay overnight in Minneapolis. You'll also refuel at each stop, except Mason City, and use the appropriate Air Route Traffic Control Centers for VFR advisories.

The reason for this itinerary is simply to review the communication procedures involved in the various situations, airports, and controlling agencies we discussed in the previous chapters. To wit:

- Kansas City Downtown lies under a TCA (Class B airspace).
- Omaha Eppley has an ARSA (Class C airspace).
- Minneapolis International is a TCA (Class B) airport.
- Mason City is a unicom airport with a Control Zone and a weather observer, but no tower or FSS on the field.
- Newton is an uncontrolled field with only unicom.

To go through the entire flight planning process is beyond the scope of this book. Let's assume that all preliminaries have been completed, including weight-and-balance calculations and filing the flight plan.

RECORDING THE FREQUENCIES

As part of the preflight planning, write down in sequence the known or probable frequencies you will use. Some, particularly Center's, might differ from what you expect, but at least you'll be ready for the majority that come into play.

I suggest you do not record all the frequencies on one piece of paper for a flight like this with five different legs. Enter the frequencies to Omaha on one sheet, those from Omaha to Minneapolis on another, and so on (FIGS. 14-2 through 14-6). Then number each sheet in sequence. As you leave one frequency and progress to the next, draw a line through (but don't obliterate) the one you have just left. This preliminary recording and progressive deleting provides cockpit organization and minimizes some of the confusion that is often the bane of the private pilot.

Two other suggestions: On each segment page, provide space for the ATIS information at the destination airport. When you're on the ground before departure, this isn't so important because you can listen to the local information as many times as necessary. In the air, however, it's another matter. Center has handed you off to Approach, but before contacting Approach, you should have monitored the ATIS (either over your second radio, or over the VOR frequency, if available)—which means that you don't have a lot of time to absorb the data being transmitted. Approach is expecting to hear from you rather promptly after the handoff.

Consequently, to expedite matters, line out a box on the flight segment page so that you can record the crucial information: 1) the phonetic alphabet; 2) the sky or ceiling; 3) visibility; 4) temperature; 5) dewpoint; 6) altimeter setting; 7) runway in use; and 8) other important information that might be included. Now, when you call

Approach, you can advise the controller that you "have Charlie," or whatever, and be sure that you have it accurately. Memories do fail us.

Second, sketch in on the same segment page a rough diagram of the destination airport runways, including the distance and direction from town, field elevation, and pattern altitude. If you want, you can add the taxiways and building locations. *AOPA's Aviation USA*, published by the Aircraft Owners and Pilots Association, provides diagrams and data of more than 13,000 airports in the United States and its possessions. It's an excellent source for determining the layout and runway data of whatever airport you have in mind. Similar diagrams are found on state aeronautical charts and on instrument approach charts.

The purpose of the sketch is apparent: it minimizes mental or spatial confusion when going into a strange airport for the first time. Just as important, it can reduce the number of questions or inquiries you might have to make of Approach, the Tower, or Ground Control.

The examples (FIGS. 14-2 through 14-6) aren't very fancy, but that's not the point. Practicality is the objective. (Just don't rely on the frequencies cited as being current. They do change.)

THE FLIGHT AND RADIO CONTACTS

Equipped with the necessary charts—sectionals as well as the Enroute Low Altitude charts (ELAC)—the frequencies recorded, the flight planned, and the flight plan filed, you're ready to depart.

Kansas City to Omaha

With the engine started, the first order of business is to tune to the Downtown Airport ATIS on 124.6. Put this in the first radio and set up Ground Control, 121.9, on the second.

> **ATIS:** *This is Kansas City Downtown Airport Information Delta. One six four five Zulu weather. Five thousand scattered, measured ceiling ten thousand broken, visibility eight. Temperature seven eight, dewpoint five five, wind one seven zero at ten, altimeter two niner niner eight. ILS Runway One Niner in use, land and depart Runway One Niner. Advise you have Delta.*

Now change the first radio to the FSS frequency of 122.6, which is the frequency that Flight Service gave you to open your flight plan. Put the transponder on STANDBY and call Ground Control on the second radio:

> **You:** Downtown Ground, Cherokee One Four Six One Tango at the Flying Service, VFR Omaha with Information Delta.
>
> **GC:** *Cherokee One Four Six One Tango, taxi to Runway One Niner.*
>
> **You:** Roger. Taxi to One Niner, Cherokee Six One Tango.

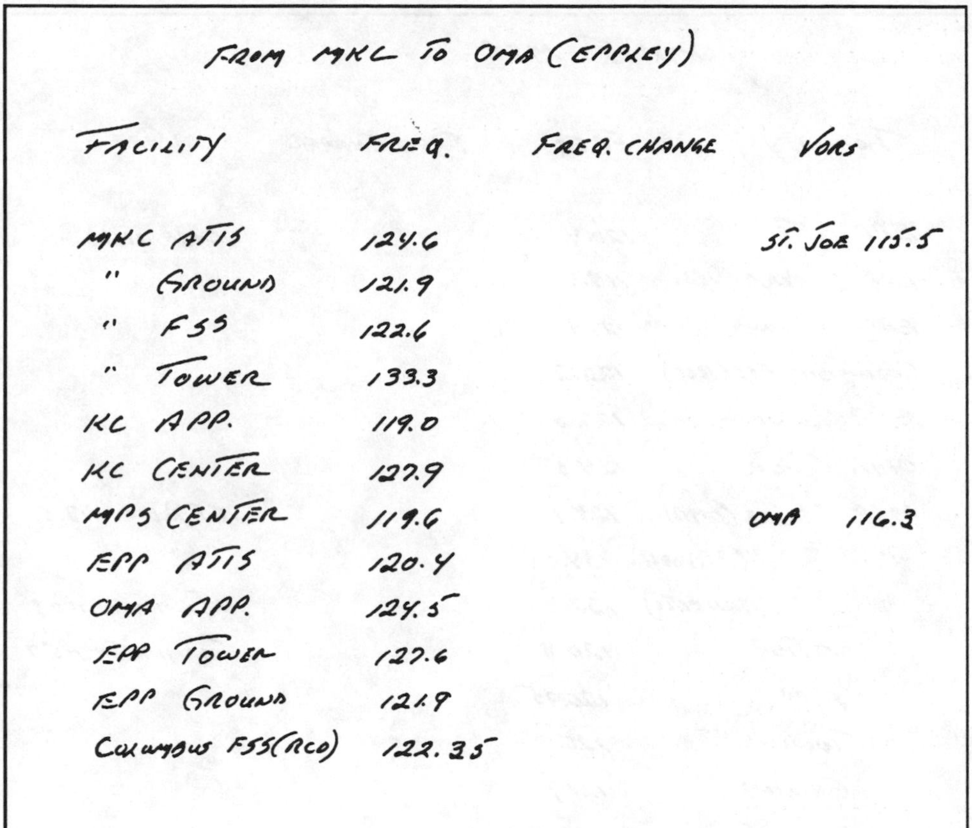

FROM MKC TO OMA (EPPLEY)

FACILITY	FREQ.	FREQ. CHANGE	VORS
MKC ATIS	124.6		ST. JOE 115.5
" GROUND	121.9		
" FSS	122.6		
" TOWER	133.3		
KC APP.	119.0		
KC CENTER	127.9		
MPS CENTER	119.6		OMA 116.3
EPP ATIS	120.4		
OMA APP.	124.5		
EPP TOWER	127.6		
EPP GROUND	121.9		
COLUMBUS FSS(RCO)	122.35		

EPP ATIS

PHONETIC _____ DEN A. _____
SKY _____ ALTIM _____
VIS _____ RWY _____
TEMP _____ OTHER _____

EPP

14R
14L 17
32R
32L
35

3 NE
ELEV.: 983
≈ 1   PATTERN: 1983

Fig. 14-2. *The flight planning notes from MKC to OMA (Eppley).*

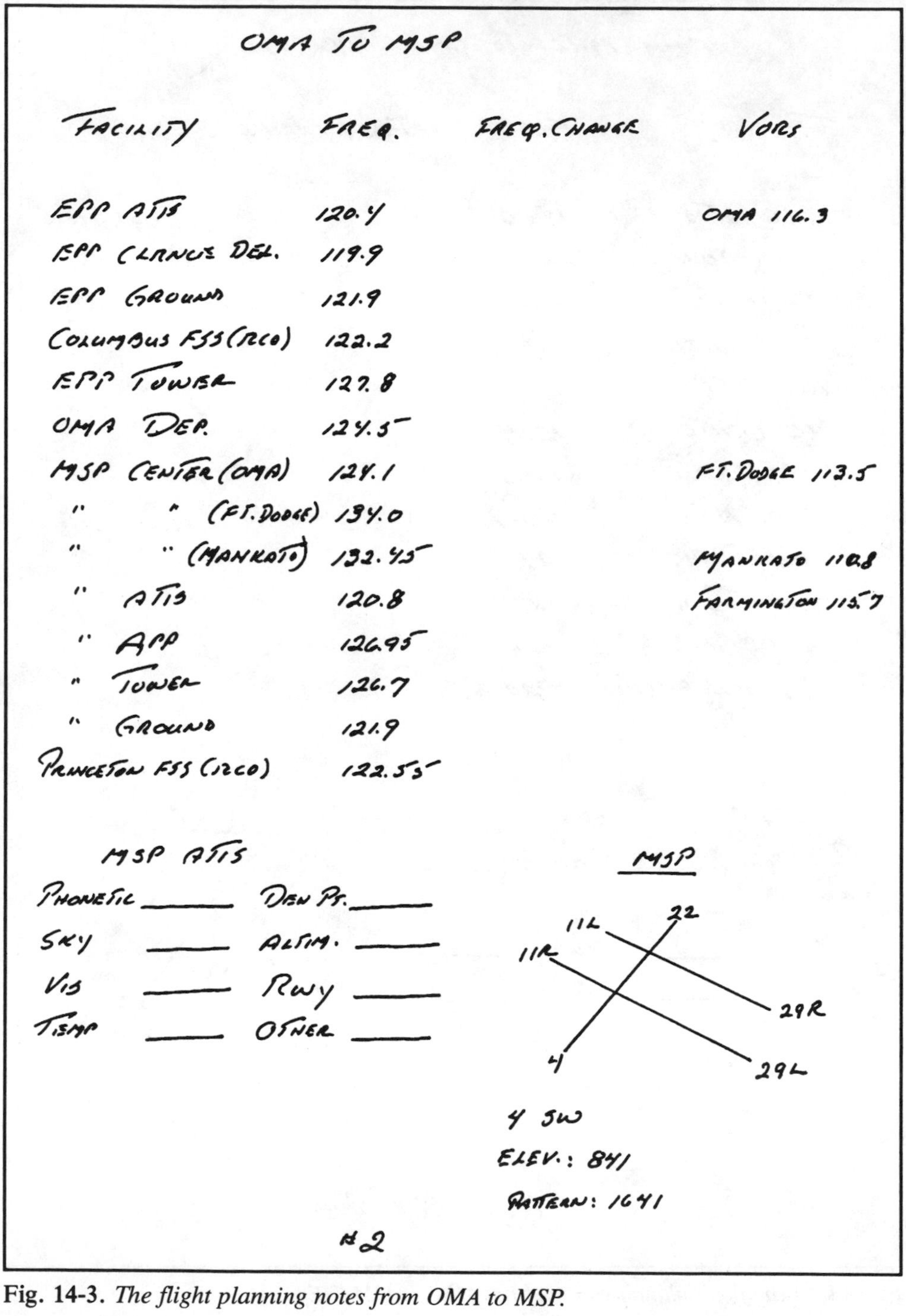

OMA TO MSP

FACILITY	FREQ.	FREQ. CHANGE	VORS
EPP ATIS	120.4		OMA 116.3
EPP CLRNCE DEL.	119.9		
EPP GROUND	121.9		
COLUMBUS FSS (RCO)	122.2		
EPP TOWER	127.8		
OMA DEP.	124.5		
MSP CENTER (OMA)	124.1		FT. DODGE 113.5
" " (FT. DODGE)	134.0		
" " (MANKATO)	132.45		MANKATO 112.8
" ATIS	120.8		FARMINGTON 115.7
" APP	126.95		
" TOWER	126.7		
" GROUND	121.9		
PRINCETON FSS (RCO)	122.55		

MSP ATIS

PHONETIC _____ DEN PT. _____
SKY _____ ALTIM. _____
VIS _____ RWY _____
TEMP _____ OTHER _____

MSP

11L
22
11R

29R

4
29L

4 SW
ELEV.: 841
PATTERN: 1641

#2

Fig. 14-3. *The flight planning notes from OMA to MSP.*

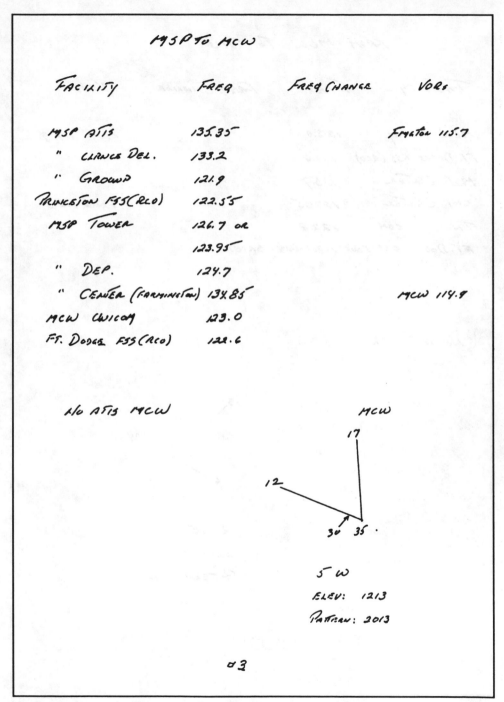

Fig. 14-4. *The flight planning notes from MSP to MCW.*

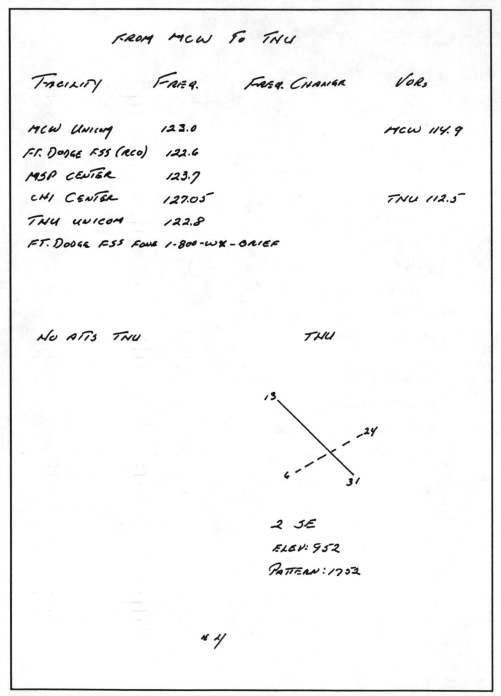

FROM MCW TO TNU

FACILITY	FREQ.	FREQ. CHANGE	VORs
MCW UNICOM	123.0		MCW 114.9
FT. DODGE FSS (RCO)	122.6		
MSP CENTER	123.7		
CHI CENTER	127.05		TNU 112.5
TNU UNICOM	122.8		
FT. DODGE FSS FONE	1-800-WX-BRIEF		

NO ATIS TNU   TNU

2 SE
ELEV: 952
PATTERN: 1752

Fig. 14-5. *The flight planning notes from MCW to TNU.*

FROM TNU TO MKC

FACILITY	FREQ	FREQ CHANGE	VORS
TNU UNICOM	122.8		TNU 112.5
FT. DODGE FSS	122.1(T); 112.5(R)		
DSM APP.	118.6 (ALL SECTORS)		DSM 114.1
MSP CENTER (OSM)	126.65		LMN 116.7
K.C. CENTER (STJOE)	127.9		MKC 112.6
MKC ATIS	124.6		
K.C. APP	119.0		
MKC TOWER	133.3		
" GROUND	121.9		
COLUMBIA FSS (RCO)	122.6		

MKC ATIS

PHONETIC ____ DEW PT. ____
SKY ____ ALTIM. ____
VIS ____ RWY ____
TEMP ____ OTHER ____

MKC
(HOME AIRPORT. NO
DIAGRAM NECESSARY)

#5

Fig. 14-6. *The flight planning notes from TNU to MKC.*

Stay on the Ground frequency. Clearance to taxi doesn't mean that the controller might not have subsequent instructions for you:

GC: *Cherokee Six One Tango, give way to the Baron taxiing south.*

You: Wilco, Cherokee Six One Tango.

After completing the pretakeoff check, either call Ground and advise them that you're leaving the frequency momentarily to go to Flight Service, or, if there's no traffic behind you that would be delayed while you make that contact, taxi to the hold line and brake to a stop. Now change radio 2 to the tower frequency of 133.3, switch to radio 1, which is already tuned to Flight Service, and open the flight plan:

You: Columbia Radio, Cherokee One Four Six One Tango on one two two point six, Kansas City.

FSS: *Cherokee One Four Six One Tango, Columbia Radio.*

You: Would you please open my VFR flight plan to Omaha Eppley at this time?

FSS: *Cherokee Six One Tango, Roger. We'll open your flight plan at five five.*

You: Roger, thank you. Cherokee Six One Tango.

Two points to remember:

1. Be sure to add five minutes, even ten, to your flight plan arrival time in case your departure is delayed.

2. Opening a flight plan while still on the ground is possible only when there is an FSS, RCO, or VOR voice facility on the field.

As your course to Omaha is northwesterly, the most direct routing is through the Kansas City TCA, so change radio 1 to Approach Control on 119.0. (Remember that you're under the TCA and must contact Approach to enter it.) With radio 1 set up, the next call is to the tower on radio 2:

You: Downtown Tower, Cherokee One Four Six One Tango ready for takeoff with north departure.

Twr: *Cherokee One Four Six One Tango, hold short. Landing traffic.*

You: Cherokee Six One Tango holding short.

Twr: *Cherokee Six One Tango, cleared for takeoff. Left turn for north departure approved. Remain clear of the final approach course. Contact Approach when airborne.*

You: Will do. Cherokee Six One Tango. [Now switch the transponder to ALT.]

Even if the tower has advised you to contact Approach after takeoff, it doesn't hurt to request the frequency-change approval or to inform the tower that you're about to make the change. The controller might have reasons for wanting you to stay with him for a few minutes. The next call then would be:

200

You: Tower, Cherokee Six One Tango requests frequency change [or "going to Approach"].

Twr: *Cherokee Six One Tango. Frequency change approved.*

You: Cherokee Six One Tango. Good day.

Go now to radio 1 and call Approach on 119.0:

You: Kansas City Approach, Cherokee One Four Six One Tango.

App: *Cherokee One Four Six One Tango, Kansas City Approach, go ahead.*

You: Cherokee One Four Six One Tango is off Downtown at two thousand, requesting six thousand five hundred to Omaha, and would like clearance through the TCA.

App: *Cherokee Six One Tango, squawk zero two five two and ident. Remain clear of the TCA.*

You: Cherokee Six One Tango squawking zero two five two.

Remember that you have not yet been cleared into the TCA, so stay below the 2400-foot floor until you hear the next message:

App: *Cherokee Six One Tango, radar contact. Cleared through the TCA. Fly heading three four zero and maintain four thousand five hundred.*

You: Roger. Cleared through the TCA, heading three four zero, leaving two thousand for four thousand five hundred. Cherokee Six One Tango.

As you start your turn and begin the climb, you might hear this before actually entering the TCA:

App: *Cherokee Six One Tango, traffic at twelve o'clock, three miles, southbound. Unverified altitude two thousand niner hundred.*

You: Cherokee Six One Tango looking.

And then:

App: *Cherokee Six One Tango, traffic no longer a factor.*

You: Roger. Cherokee Six One Tango.

You: Cherokee Six One Tango, level at four thousand five hundred.

App: *Cherokee Six One Tango, Roger. Climb and maintain six thousand five hundred.*

You: Out of four thousand five hundred for six thousand five hundred. Cherokee Six One Tango.

You: Cherokee Six One Tango level at six thousand five hundred.

App: *Cherokee Six One Tango, Roger. Turn left heading three one zero. Proceed direct St. Joe when able.*

You: Roger. Three one zero on the heading. Direct St. Joe when able. Cherokee Six One Tango.

The St. Joe VOR lies about 50 nm north of Kansas City. Approach is saying that you are to tune the nav receiver to the VOR frequency of 115.5. When you have the altitude and distance to get a steady needle reading, you're cleared to proceed directly on course to St. Joe.

Other instructions or traffic alerts might follow. While you have the time, however, you should be setting up radio 2 (until now still on the tower frequency) to Kansas City Center, which you expect to be 127.9. Approach might give you a different frequency, but if not, you're ready to contact Center without delay.

As you work your way northward toward St. Joe, you reach the TCA limits:

App: *Cherokee Six One Tango, position 30 miles northwest of International, departing the TCA. Radar service terminated. Squawk one two zero zero. Frequency change approved.*

You: Cherokee Six One Tango. Can you hand us off to Center?

App: *Unable at this time, Six One Tango. Contact Center on one two seven point niner. Good day.*

You: One two zero zero and one two seven point niner. Roger. Cherokee Six One Tango.

There is no handoff, so the call to Center requires the full IPAI/DS:

You: Kansas City Center, Cherokee One Four Six One Tango.

Ctr: *Cherokee One Four Six One Tango, Kansas City Center, go ahead.*

You: Cherokee One Four Six One Tango is two zero south of St. Joe VOR, level at six thousand five hundred enroute Omaha. Squawking one two zero zero. Request VFR advisories, if possible.

Ctr: *Cherokee Six One Tango, squawk four one five zero and ident.*

You: Cherokee Six One Tango squawking four one five zero.

Ctr: *Cherokee Six One Tango. Radar contact. St. Joe altimeter two niner niner eight.*

You: Roger. Cherokee Six One Tango.

What comes over the air now depends primarily on the traffic along your line of flight. You might be told to assume a different heading; you might be alerted to the proximity of other aircraft; you might be alerted to military training flights. Or you might hear nothing.

Regardless of the communiqués, or lack thereof, you have some navigating to do, along with keeping an ear out for your call sign. You're still on course to the St. Joe VOR, with the VOR head needle centered and the DME recording the distance to the VOR, the time to the station, and your current ground speed.

As soon as you pass over St. Joe and get a "FROM" reading on the VOR, turn to a heading of 325°. Once the needle has centered itself, indicating that you're now on V159, stay tuned to that station for another 35 or 40 miles. If you have a second nav-com, you can tune in the Omaha VOR on 116.3 at any time, but at your altitude you probably won't get a very reliable "TO" indication until you're within a 60-or 70-mile range of the station. So maintain the outbound course from St. Joe with the "FROM" reading on one VOR, and then when the other VOR needle settles down to a steady centered position with a "TO" reading, rely on it to lead you to the Omaha station.

Meanwhile, don't get too enthralled with VOR-to-VOR navigating alone. Something might go wrong, so it's always wise to check your progress against the route you've laid out on the sectional. If you should lose the nav portion of the radio, or if the VOR suddenly had a mechanical, it could be rather important to know where you are along the route. Things electronic are great, but they're not infallible.

Eventually, as you move along V159, Kansas City Center comes on the air:

Ctr: *Cherokee Six One Tango, contact Minneapolis Center now on one one niner point six. Good day.*

You: One one niner point six. Thank you for your help. Cherokee Six One Tango.

No comment here about radar service being terminated, so this is a handoff by Kansas City to Minneapolis. If you've already changed radio 1 from Kansas City Approach to Minneapolis, all you have to do now is go back to radio 1 and introduce yourself:

You: Minneapolis Center, Cherokee One Four Six One Tango with you, level at six thousand five hundred.

Ctr: *Cherokee Six One Tango, Roger. Altimeter three zero zero two.*

Unless you are told otherwise, maintain your present heading and altitude. Again, there might or might not be instructions or advice for you. Eventually, however, as you near the Omaha ARSA, Center will come back on:

Ctr: *Cherokee Six One Tango, position 30 miles south of Eppley. Contact Omaha Approach on one two four point five. Good day.*

You: Roger. One two four point five. Cherokee Six One Tango. Good day.

This, too, is a handoff, so tune in to Approach. Before making the call, however, if you haven't done so already, get the ATIS on 120.4 for the current Eppley information. Now you're ready to contact Approach:

You: Omaha Approach, Cherokee One Four Six One Tango is with you, level at six thousand five hundred with Information Echo.

App: *Cherokee Six One Tango, Roger. Maintain present heading and altitude.*

As you enter the ARSA and draw close to the field, you're likely to be given a variety of instructions that will sequence you into the existing traffic flow. Whatever the messages, be sure to acknowledge and repeat them in an abbreviated form:

App: *Cherokee Six One Tango, turn right heading three five zero. Descend and maintain four thousand.*

You: Right to three five zero. Leaving six thousand five hundred for four thousand. Cherokee Six One Tango.

In another few minutes, you'll hear from Approach again:

App: *Cherokee Six One Tango, Eppley is at twelve o'clock, six miles. Contact Eppley Tower on one two seven point six.*

You: Roger. One two seven point six—and we have the field in sight. Cherokee Six One Tango. Good day.

Another handoff:

You: Eppley Tower, Cherokee One Four Six One Tango with you, level at four thousand.

Twr: *Cherokee Six One Tango, enter left downwind for Runway One Four left. Sequence later.*

Two points here:

1. "Sequence later" simply means the tower will advise you in due time whether you're cleared to land, are "number two following a Cessna on downwind," "number four following the Duchess," or whatever. Just remember to tell the tower that you "have the Cessna" or "negative contact on the Duchess" or "have the traffic." Always keep the tower informed—don't leave them in the dark.

2. You'll note that you've used the same squawk from Kansas City Center through Minneapolis Center, Omaha Approach, and Eppley Tower. No controlling agency has asked you to change. This is not always the case. Any one of them could have requested a different squawk and an ident. Leave the transponder on the current squawk until directed otherwise.

With a little time available now, dial in Eppley's Ground frequency, 121.9, on radio 1. Then, in due course, you'll hear on radio 2:

Twr: *Cherokee Six One Tango, cleared to land.*

You: Roger, cleared to land. Cherokee Six One Tango.

During the landing rollout, the tower makes its final contact with you:

Twr: *Cherokee Six One Tango, contact Ground point niner clear of the runway.*

You: Cherokee Six One Tango, wilco.

Suppose, however, that you're not sure whether to make a left or right turnoff.

You want to go to a Phillips dealer but don't know where one is located. Despite the uncertainty, don't tie up the tower frequency by asking the controller for directions. Let Ground do this for you, even if it means being cleared back across the active runway because you turned right instead of left, or vice versa. Merely acknowledge the tower's instructions and go to Ground's frequency:

You: Eppley Ground, Cherokee One Four Six One Tango Clear of One Four Left. Request progressive taxi to the Phillips dealer.

GC: *Cherokee One Four Six One Tango, turn right next taxiway. Taxi to One Four Right and hold for departing traffic.*

You: Roger, right and hold at One Four. Six One Tango.

After a minute or two:

GC: *Cherokee Six One Tango, Clear to cross One Four Right. Phillips is to your right at the Sky Harbor FBO.*

You: Roger, Ground. Clear to cross One Four—and we have the FBO. Six One Tango.

You're at the ramp, with the engine cut. Now don't forget to call Flight Service by phone or radio to close out your flight plan.

Omaha to Minneapolis

After an hour on the ground for refueling, a bite to eat, and a call to the Columbus, Nebraska, AFSS for a weather briefing and filing the flight plan, you're ready to go again, with friends and baggage on board. The only new element is the need to call Clearance Delivery for initial VFR instructions within the ARSA. As you've already determined, that frequency is 119.9, and you were told to contact Flight Service on the Omaha RCO frequency of 122.35. That's a change from the 122.2 you had listed on the frequency chart you prepared back in Kansas City (FIG. 14-3). So cross off 122.2 on that chart and enter 122.35 under the "Changes" column.

As usual, the communication chain begins by monitoring ATIS: "This is Eppley Information Gulf. One niner four five Zulu weather. Eight thousand scattered, visibility five, haze and smoke. Temperature eight five, dewpoint six two. Wind one four zero at one five. Altimeter three zero two five. ILS Runway One Four Right in use, land and depart Runway One Four Right. Advise you have Gulf."

Next, the call to Clearance (CD):

You: Eppley Clearance, Cherokee One Four Six One Tango.

CD: *Cherokee One Four Six One Tango, Eppley Clearance.*

You: Cherokee One Four Six One Tango will be departing Eppley, VFR northeast for Minneapolis at seven thousand five hundred.

CD: *Cherokee Six One Tango, Roger. Turn left heading zero four five after*

> *departure. Climb and maintain three thousand. Squawk two four four
> zero. Departure frequency one two four point five.*

You: Roger. Zero four five on the heading, maintain three thousand, two four four
zero, and one two four point five. Cherokee Six One Tango.

CD: *Cherokee Six One Tango, readback correct.*

You: Cherokee Six One Tango.

Before calling Ground on radio 2, dial out the Clearance frequency in radio 1 and
replace it with Flight Service's—122.35. Also, put the transponder on STANDBY.

You: Eppley Ground, Cherokee One Four Six One Tango at Sky Harbor with Information Gulf. Ready to taxi with clearance.

GC: *Cherokee Six One Tango, taxi to Runway One Four Right for intersection
departure.*

You: Ground, Cherokee Six One Tango would like full length.

GC: *Cherokee Six One Tango, roger. Full length approved.*

You: Cherokee Six One Tango.

The pretakeoff check completed, the next call is to Flight Service after advising
Ground that you were going to change frequencies momentarily.

You: Columbus Radio, Cherokee One Four Six One Tango on one two two point
three five, Eppley.

FSS: *Cherokee One Four Six One Tango, Columbus Radio.*

You: Columbus, Cherokee Six One Tango. Would you open my VFR flight plan to
Minneapolis International at this time?

FSS: *Cherokee Six One Tango, Roger. We'll open your flight plan at two five.*

You: Roger, thank you. Cherokee Six One Tango.

That done, advise Ground Control that you're back with them. Then taxi to the hold
and call the tower:

You: Eppley Tower, Cherokee One Four Six One Tango ready for takeoff with
northeast departure.

Twr: *Cherokee Six One Tango, taxi into position and hold.*

You: Position and hold. Cherokee Six One Tango. [Switch the transponder from
STANDBY to ALT as you're taxiing to the runway.]

Twr: *Cherokee Six One Tango, cleared for takeoff.*

You: Roger, cleared for takeoff. Cherokee Six One Tango. Is northeast departure
approved?

Twr: *Affirmative, northeast departure approved.*

You: Roger. Cherokee Six One Tango.

When airborne, request the frequency change to Departure—if the tower has not already advised you to do so:

You: Tower, Cherokee Six One Tango requests frequency change to Departure.

Twr: *Cherokee Six One Tango, frequency change approved.*

You: Roger. Cherokee Six One Tango. Good day.

To Departure Control:

You: Omaha Departure, Cherokee One Four Six One Tango is with you out of one thousand eight hundred for three thousand. Request seven thousand five hundred.

Dep: *Cherokee Six One Tango, radar contact. Maintain present heading. Report level at three thousand.*

You: Roger, report three thousand, Cherokee Six One Tango.

You: Cherokee Six One Tango level at three thousand.

Dep: *Cherokee Six One Tango, Roger. Turn left heading zero three zero.*

You: Left to zero three zero. Cherokee Six One Tango.

Dep: *Cherokee Six One Tango, climb and maintain seven thousand five hundred.*

You: Roger. Out of three thousand for seven thousand five hundred. Cherokee Six One Tango.

You: Cherokee Six One Tango level at seven thousand five hundred.

Dep: *Cherokee Six One Tango, Roger. Proceed on course and contact Minneapolis Center on one two four point one.* [An unsolicited handoff for advisories, although rare, is possible.]

You: One two four point one. Thank you for your help. Cherokee Six One Tango.

To Center:

You: Minneapolis Center, Cherokee One Four Six One Tango with you, level at seven thousand five hundred.

Ctr: *Cherokee Six One Tango, Roger. Radar contact. Eppley altimeter Three Zero Two Five.*

You: Roger. Cherokee Six One Tango. Three Zero Two Five.

Down the road apiece, the summer turbulence at 7,500 feet is getting a bit too much for one of your passengers, so you decide to climb to 9,500. But first:

You: Minneapolis Center, Cherokee Six One Tango is out of seven thousand five hundred for niner thousand five hundred due to turbulence.

Ctr: *Cherokee Six One Tango, Roger. Report level at niner thousand five hundred.*

You: Wilco, Cherokee Six One Tango.

You: Cherokee Six One Tango level at niner thousand five hundred.

Ctr: Cherokee Six One Tango, Roger.

Just remember that on a VFR flight plan outside of a terminal area, you're free to deviate from existing headings and altitudes. But since you're asking Center for enroute advisories, don't make changes without advising the controller of your intentions. Although he can spot heading changes on the screen (and altitude changes, if you're equipped with a Mode C transponder), be sure to warn him in advance.

As you proceed toward Fort Dodge on V138, Center comes on again:

Ctr: Cherokee Six One Tango, contact Minneapolis Center now on one three four point zero. [This is the Fort Dodge remote outlet.]

You: Roger. One three four point zero. Cherokee Six One Tango. Good day.

Change the frequency accordingly and reestablish yourself with Center:

You: Minneapolis Center, Cherokee One Four Six One Tango with you level at niner thousand five hundred.

Ctr: Cherokee Six One Tango, Roger. Fort Dodge altimeter three zero one six. [Plus any instructions or traffic advisories the controller might have for you.]

Crossing the Fort Dodge VOR, you make a time check against your flight plan and find that you're running about 20 minutes behind schedule, which is quite a difference for the 110-mile flight from Eppley. Either the forecast winds have changed or those at the 9,500-foot altitude are from a different direction or velocity. Whatever the case, and because you intend to stay at the same altitude and generally northeast direction, it's fair to assume that your arrival in Minneapolis will be later than planned. With another 150 miles to go and at the present ground speed, you could easily be 45 minutes to an hour later than the flight plan ETA.

Before taking any arbitrary action, you decide that more information about the winds is in order, thus a call to Flight Service. First, however, advise Center if you're temporarily going to leave the frequency:

You: Center, Cherokee Six One Tango is leaving you temporarily to go to Flight Service.

Ctr: Cherokee Six One Tango, Roger. Advise when you're back with me.

You: Will do. Cherokee Six One Tango.

The FSS call is made on 122.3, the Fort Dodge transmit/receive frequency: (*Note:* the Fort Dodge AFSS does not have Flight Watch service—which is why you make the call on 122.3 rather than 122.0.)

You: Fort Dodge Radio, Cherokee One Four Six One Tango on one two two point three.

FSS: *Cherokee One Four Six One Tango, Fort Dodge Radio, go ahead.*

You: Fort Dodge, Cherokee Six One Tango is just north of the Fort Dodge VOR at niner thousand five hundred. Request winds at niner thousand feet.

FSS: *Cherokee Six One Tango, Roger. Stand by.*

FSS: *Cherokee Six One Tango, winds at niner thousand are three five zero at four zero. Fort Dodge altimeter two niner two five.*

You: Three five zero at four zero and two niner two five. Thank you. Cherokee Six One Tango.

Without going through the mechanics of recomputing your ground speed and ETA based on this information, let's just say that you determine that your arrival will be 48 minutes later than your flight plan forecast. This conclusion warrants another call to Flight Service:

You: Fort Dodge Radio, Cherokee One Four Six One Tango on one two two point three.

FSS: *Cherokee One Four Six One Tango, Fort Dodge Radio, go ahead.*

You: Cherokee One Four Six One Tango is one zero north of the Fort Dodge VOR on VFR flight plan to Minneapolis International with a one six four five local ETA. Would like to extend the ETA to one seven three zero.

FSS: *Cherokee Six One Tango, Roger. Will extend your ETA to one seven three zero local. Fort Dodge altimeter two niner two five.*

You: Roger, thank you. Cherokee Six One Tango.

The next move is to go back to Center and reestablish contact:

You: Center, Cherokee Six One Tango is back with you.

Ctr: *Cherokee Six One Tango, Roger.*

With that taken care of, you can rest more easily. You have until 1800 before Flight Service will start asking questions as to your whereabouts. You've also allowed yourself an additional 27 minutes as an extra cushion.

Moving northward along V456, you'll be approaching another Center change point—this time to the remoted Mankato site. Somewhere near Mankato, you get this call:

Ctr: *Cherokee Six One Tango, contact Minneapolis Center now on one three two point four five.*

You: One three two point four five. Cherokee Six One Tango. Good day.

You: Minneapolis Center, Cherokee One Four Six One Tango with you, level at niner thousand five hundred.

Ctr: *Cherokee Six One Tango, Roger. Mankato altimeter two niner two three.*

You: Roger, two niner two three. Cherokee Six One Tango.

Immediately after passing the Mankato VOR, you leave the airway and turn to about 30 degrees in the direction of the Farmington VOR. Because you're only 50 miles from the airport at Mankato, however, it's almost certain that Center will turn you over to Approach and Approach will vector you to the airport ATA. That means you might not come anywhere near Farmington. Meanwhile, the closer you edge toward the Minneapolis TCA (Class B), the greater the likelihood that Center will offer traffic advisories and possible heading changes. There might be none, but moving into a busy terminal environment increases the possibility.

About 20 miles out from the TCA limit, Center will conclude its radar surveillance. Perhaps the controller will hand you off to Approach. If so, the usual "with you," plus your altitude, is all that's necessary. If there is no handoff, though, you'll have to give the IPAI/DS—which means, among other things, knowing your position or approximate distance from the airport.

In this case; let's assume that Center is getting busy and doesn't have time to contact Approach. The controller comes on with:

Ctr: *Cherokee One Four Six One Tango, radar service terminated. Squawk one two zero zero. Contact Minneapolis Approach on one two six point niner five. Descend your discretion.*

You: One two six point niner five, Will do. Cherokee Six One Tango.

"Descend your discretion" simply means what it says: You are clear to begin losing altitude whenever you wish and at whatever rate you wish. With that approval, you throttle back so that you'll be down to about 4,000 feet by the time you near the 30-mile TCA veil.

There's one more thing: Before calling Approach, dial in 120.8 for the current ATIS. Then go to 126.95, which is the Approach frequency for aircraft arriving from the south and west, and make the introductory call:

You: Minneapolis Approach, Cherokee One Four Six One Tango.

App: *Cherokee One Four Six One Tango, Minneapolis Approach.*

You: Approach, Cherokee Six One Tango is thirty five southwest over Le Center at five thousand three hundred descending for landing International and squawking one two zero zero with Delta.

App: *Cherokee Six One Tango, squawk four one two two and ident. Remain clear of the TCA (Class B airspace).*

You: Roger, four one two two, and remaining clear. Cherokee Six One Tango.

App: *Cherokee One Four Six One Tango, radar contact thirty-two miles southwest. Cleared into the TCA (or Minneapolis Class B airspace). Turn left to zero two zero, descend and maintain four thousand five hundred.*

You: Roger, Cherokee Six One Tango cleared into the TCA (or Class B airspace). Left to zero two zero and descending to four thousand five hundred.

You: Approach, Cherokee Six One Tango level at four thousand five hundred.

App: *Roger Six One Tango. Maintain heading and altitude.*

You: Will do. Six One Tango.

A short time later.

App: *Cherokee One Four Six One Tango, turn left to zero one zero and descend to two thousand five hundred.*

You: Roger, left to zero one zero and down to two thousand five hundred, Cherokee Six One Tango.

You: Approach, Cherokee Six One Tango level at two thousand five hundred.

App: *Roger, Six One Tango. Contact Minneapolis Tower on one two six point seven.*

You: One two six point seven. Will do, and thank you. Cherokee Six One Tango.

You: Minneapolis Tower, Cherokee One Four Six One Tango is with you, level at two thousand five hundred.

Twr: *Roger, Cherokee One Four Six One Tango. Descend to one thousand eight hundred for straight-in approach, landing Runway Four. Sequence later.*

You: Roger, descending to one thousand eight hundred and Runway Four. Cherokee Six One Tango.

You: Tower, Cherokee Six One Tango level at one thousand eight hundred.

Twr: *Roger, Six One Tango, you will be number two to land behind the Baron. Advise when you have traffic in sight.*

You: Number two, and will advise. Six One Tango.

You: Tower, Cherokee Six One Tango has the Baron.

Twr: *Roger Six One Tango. Thank you.*

Twr: *Cherokee Six One Tango, cleared to land Runway Four.*

You: Cherokee Six One Tango cleared to land.

When you're down:

Twr: *Cherokee Six One Tango, contact Ground point niner.*

You: Will do, tower. Six One Tango.

When you're clear of the active and at a full stop:

You: Minneapolis Ground, Cherokee One Four Six One Tango clear of Four. Taxi to Airmotive.

GC: *Roger. Cherokee One Four Six One Tango, taxi to Airmotive.*

Once again, when you're parked or in the pilot's lounge, close out the flight plan with Flight Service—which, in this case, is located in Princeton, Minnesota.

Minneapolis to Mason City

It's the next morning and you're ready to set out for Mason City, Newton, and then back to Kansas City. First the normal routines: checking the weather, filing the flight plan, and determining the FSS frequency to open the flight plan. With the engine started, monitor the Minneapolis departure ATIS on 135.35. Remember that you're in a TCA, so the first call goes to Clearance Delivery on 133.2:

You: Minneapolis Clearance, Cherokee One Four Six One Tango.

CD: *Cherokee One Four Six One Tango, Clearance.*

You: Cherokee One Four Six One Tango will be departing VFR to Mason City. Request seven thousand five hundred.

CD: *Cherokee Six One Tango, Roger. Turn right heading one seven five after departure. Climb and maintain three thousand. Squawk zero four two zero. Departure frequency one two four point seven.*

You: Right heading one seven five after departure, maintain three thousand, zero four two zero, and one two four point seven.

CD: *Cherokee Six One Tango, readback correct.*

You: Roger. Cherokee Six One Tango.

Now to Ground Control on 121.9:

You: Minneapolis Ground, Cherokee One Four Six One Tango at Airmotive with Information Echo and clearance.

GC: *Cherokee Six One Tango, taxi to Runway One One Right.*

You: Roger, One One Right, Cherokee Six One Tango.

The pretakeoff check has been completed. Before taxiing to the hold line, ask Ground for permission to switch frequencies, then call Flight Service to open the flight plan.

You: Princeton Radio, Cherokee One Four Six One Tango on one two two point five five [the RCO frequency FSS gave you when filing], Minneapolis.

FSS: *Cherokee One Four Six Tango, Princeton Radio.*

You: Roger, would you please open my flight plan to Mason City at this time?

FSS: *Cherokee Six One Tango, Roger. Will open your flight plan at one zero.*

You: Thank you. Cherokee Six One Tango.

As you move toward the hold line, you see that two aircraft are ahead of you awaiting takeoff permission. Regardless, you pull behind the second plane and call the tower on 126.7:

You: Minneapolis Tower, Cherokee One Four Six One Tango ready for takeoff, number three in sequence, south departure.

Twr: *Cherokee One Four Six One Tango, taxi around the Mooney and Skymaster. Cleared for takeoff, south departure approved.*

You: Roger, cleared for takeoff, Cherokee Six One Tango.

At this point, switch the transponder from STANDBY to ALT. When airborne, turn to your assigned heading of 175 degrees. The tower will probably authorize the frequency change to Departure, but if it doesn't, request the change:

You: Tower, Cherokee Six One Tango requests frequency change to Departure.

Twr: *Cherokee Six One Tango, frequency change approved. Good day.*

You: Roger. Cherokee Six One Tango. Good day.

Switch to your other radio, already dialed in to 124.7:

You: Minneapolis Departure, Cherokee One Four Six One Tango with you out of one thousand seven hundred for three thousand.

Dep: *Cherokee One Four Six One Tango, report level at three thousand.*

You: Wilco, Cherokee Six One Tango.

You: Cherokee Six One Tango level at three thousand.

Dep: *Cherokee Six One Tango, Roger.*

Dep: *Cherokee Six One Tango, climb and maintain four thousand five hundred.*

You: Roger. Out of three thousand for four thousand five hundred. Cherokee Six One Tango.

You: Cherokee Six One Tango level at four thousand five hundred.

Dep: *Cherokee Six One Tango, Roger. Cleared direct Farmington VOR.*

You: Roger, cleared direct Farmington. Cherokee Six One Tango.

Dep: *Cherokee Six One Tango, climb and maintain seven thousand five hundred.*

You: Roger. Out of four thousand five hundred for seven thousand five hundred. Cherokee Six One Tango.

You: Cherokee Six One Tango level at seven thousand five hundred.

Dep: *Cherokee Six One Tango. Roger. Report Farmington VOR.*

You: Roger, report Farmington. Cherokee Six One Tango.

It isn't long before the Course Direction Indicator (CDI) swings erratically and the VOR flag changes from TO to FROM. You're over the Farmington VOR. You report in and are cleared to turn to the 178-degree heading which establishes you on Victor 13. As you're still in the TCA, other instructions might be forthcoming from Departure. In a very few minutes, you'll hear something like this:

Dep: *Cherokee Six One Tango, position fifteen miles south of Farmington VOR,*

> *departing the TCA. Radar service terminated. Squawk one two zero zero.*
> *Frequency change approved. Good day.*

You: Departure, Cherokee Six One Tango. Can you hand us off to Center?

Dep: *Cherokee Six One Tango, stand by.* [Pause] *Cherokee Six One Tango,*
unable at this time. Squawk one two zero zero. Suggest you contact Min-
neapolis Center on one three four point eight five.

You: Roger. One two zero zero and one three four point eight five. Cherokee Six
One Tango.

You: Minneapolis Center, Cherokee One Four Six One Tango.

Ctr: *Cherokee One Four Six One Tango, Minneapolis Center.*

You: Cherokee One Four Six One Tango is fifteen south of the Farmington VOR at
seven thousand five hundred, VOR to Mason City, squawking one two zero
zero. Request VFR advisories, if possible.

This time, the press of traffic in and out of the Minneapolis area is of such density
that Center can't accept your request:

Ctr: *Cherokee One Four Six One Tango, unable at this time. Suggest you moni-*
tor this frequency.

You: Roger, Center. Understand. Cherokee Six One Tango.

With only 80 miles or so to go, this isn't much of a problem. However, it does
mean that the need for constant sky-scanning is more important than ever. If Center is
too busy to give you advisories, you can be reasonably certain that there's a fair
amount of activity in the surrounding area. Maximum alertness is in order.

Approaching Mason City, note that the airport has no tower or FSS but there is a
Control Zone, which tells you that a qualified weather observer is located on or near
the field. Contact with the Fort Dodge AFSS is via the 122.6 RCO, while unicom can
be reached on 123.0.

Also note that the Mason City VORTAC is on the southern fringe of the Control
Zone, or 4 nautical miles from the airport. If you're tracking the VOR into Mason
City from the north, as you are, and have the DME tuned to the VOR frequency, you
can quickly determine your distance from the field: the mileage to the VOR minus 4
nm.

Preparing for the landing, you begin descending about 20 miles out, and shortly
thereafter call unicom on 123.0:

You: Mason City Unicom, Cherokee One Four Six One Tango.

Uni: *Cherokee One Four Six One Tango, Mason City Unicom.*

You: Unicom, Cherokee One Four Six One Tango is fifteen north at five thousand
three hundred descending for landing. Request field advisory.

Uni: *Six One Tango, wind is two zero zero at one zero, altimeter two niner four*
five. Favored runway is One Seven. No reported traffic.

You: Roger, thank you. Six One Tango.

From this point on, remember to address all flight operation reports to "Mason City Traffic," not unicom. If, however, your remaining passenger wants a cab or a telephone call made, or you have a request of a nonoperational nature, the message is addressed to "Mason City Unicom."

Down at pattern altitude and nearing the Control Zone, you make the first traffic call, even though there is "no reported" traffic and you've heard no one on the air.

You: Mason City Traffic, Cherokee One Four Six One Tango is six miles north at two thousand for straight-in approach Runway One Seven, full stop, Mason City.

You: Mason City Traffic, Cherokee Six One Tango on two mile final for One Seven, full stop, Mason City.

When down and clear of the runway:

You: Mason City Traffic, Cherokee Six One Tango clear of One Seven, taxiing to the ramp, Mason City.

At the ramp you decide to close the flight plan by radio rather than by telephone. Using the RCO you call the AFSS:

You: Fort Dodge Radio, Cherokee One Four Six One Tango on one two two point six, Mason City.

FSS: *Cherokee One Four Six One Tango, Fort Dodge Radio, go ahead.*

You: Cherokee Six One Tango is on the ramp at Mason City. Would you close out my VFR flight plan from Minneapolis at this time?

FSS: *Cherokee Six One Tango, Roger. Will close out your flight plan at five five.*

You: Thank you. Cherokee Six One Tango.

Mason City to Newton

The flight plan to Newton, 90 miles away, has been filed, you know the weather, and it's departure time again. Tune to 123.0 again and announce your initial intentions:

You: Mason City Traffic, Cherokee One Four Six One Tango at the terminal, taxiing to Runway One Seven, Mason City.

After the pretakeoff check, with the second radio already turned to 122.6, open the flight plan:

You: Fort Dodge Radio, Cherokee One Four Six One Tango on one two two point six, Mason City.

FSS: *Cherokee One Four Six One Tango, Fort Dodge Radio, go ahead.*

> **You:** Fort Dodge, would you please open my flight plan to Newton at this time?
>
> **FSS:** *Cherokee Six One Tango, Roger. Will open your flight plan to Newton at two zero.*
>
> **You:** Roger. Thank you. Cherokee Six One Tango.

Now go back to 123.0:

> **You:** Mason City Traffic, Cherokee Six One Tango taking One Seven, straight-out departure, Mason City.

When clear of the Control Zone:

> **You:** Mason City Traffic, Cherokee Six One Tango is clear of the area to the south, Mason City.

Following this call, you request Minneapolis Center to give you VFR advisories. Center in this location is remoted to Mason City on 127.3:

> **You:** Minneapolis Center, Cherokee One Four Six One Tango.
>
> **Ctr:** *Cherokee One Four Six One Tango, Minneapolis Center, go ahead.*
>
> **You:** Center Cherokee One Four Six One Tango is off Mason City at four thousand five hundred, climbing to seven thousand five hundred, enroute Newton, and squawking one two zero zero. Request VFR advisories.
>
> **Ctr:** *Cherokee Six One Tango, squawk two five two five and ident.*
>
> **You:** Cherokee Six One Tango, two five two five.
>
> **Ctr:** *Cherokee Six One Tango, radar contact. Traffic at ten o'clock, four miles, southbound. Altitude unknown.*
>
> **You:** Cherokee Six One Tango is looking.

A minute or so later, you spot the target a little above you at the eleven o'clock position:

> **You:** Cherokee Six One Tango has the traffic.
>
> **Ctr:** *Cherokee Six One Tango, Roger.*

When at your altitude:

> **You:** Cherokee Six One Tango level at seven thousand five hundred.
>
> **Ctr:** *Cherokee Six One Tango, Roger.*

Very shortly, according to the Enroute Low Altitude Chart (ELAC), you'll be leaving the airspace of Minneapolis Center and entering that controlled by Chicago. You can't be sure, but you'll probably be asked to change to the Des Moines remote outlet on 127.05. Assuming that will be the frequency, dial it in so that you'll be prepared. After a few more minutes, Center comes on:

> **Ctr:** *Cherokee Six One Tango, contact Chicago Center now on one two seven point zero five. Good day.*

You: Roger, one two seven point zero five. Thank you for your help. Cherokee Six One Tango.

You: Chicago Center, Cherokee One Four Six One Tango is with you, level at seven thousand five hundred.

Ctr: *Cherokee Six One Tango, radar contact. Des Moines altimeter two niner niner eight.*

Nearing Newton, you'll hear something like this:

Ctr: *Cherokee Six One Tango, position one five miles north of the Newton VOR. Radar service terminated. Squawk one two zero zero. Frequency change approved. Good day.*

You: Roger. One two zero zero. Thank you for your help. Cherokee Six One Tango.

Newton is uncontrolled, with only unicom on 122.8. About 10 miles out, you announce your presence:

You: Newton Unicom, Cherokee One Four Six One Tango is five north of Newton VOR. Request airport advisory.

Uni: *Cherokee One Four Six One Tango, Newton Unicom. Wind is two one zero at one five. Altimeter three zero one five. Runway One Three in use. Three Cessnas reported in the pattern.*

You: Roger. Cherokee Six One Tango.

You: Newton Traffic, Cherokee One Four Six One Tango, eight miles northwest at four thousand. Descending for straight-in approach, Runway One Three, full stop, Newton.

You continue to descend and get lined up with the runway. Just as you're about to announce your position on the 3-mile final, one of the Cessnas comes on the air: "Cessna Six Four Foxtrot turning base, touch-and-go, One Three, Newton."

You spot the Cessna in his turn to base and realize that the two of you are going to have a fairly close final together if things keep on as they are. You have a choice: "fly" the final with a series of S-turns, or do a 360. Of the two, the latter seems the wiser move:

You: Newton Traffic, Cherokee Six One Tango is on a 3-mile final but will do a three sixty to give way to the Cessna on base, Newton.

The Cessna might or might not thank you for your courtesy. Regardless, you complete the maneuver and line up for One Three again:

You: Newton Traffic, Cherokee Six One Tango on three-mile final for Runway One Three, full stop, Newton.

When down and clear of the runway:

You: Newton Traffic, Cherokee Six One Tango clear of Runway One Three, taxiing to the terminal, Newton.

When in the terminal, don't forget to cancel the flight plan. At Newton, this has to be done by phone to the FSS in Fort Dodge on 1-800-WX-BRIEF.

Newton to Kansas City

With the remaining passenger dropped off and full fuel tanks, you're ready for the last leg back to Kansas City. But first comes another call to Flight Service for a weather check and flight plan filing. Local conditions are determined from the unicom operator (or review the winds, etc., yourself if he's out gassing an airplane).

The radio contacts will first be to local traffic and then to Flight Service, which you won't be able to reach until you have some altitude. Heading southwest to the Des Moines VOR, you'll be well above the 5,000 foot msl ceiling of the Des Moines ARSA, so calling Approach isn't necessary. From the VOR southbound on V13, however, you'll be in the ARSA's outer area, which rises to approximately 12,000 feet msl, for a few minutes. Even though the weather is fine, you conclude that monitoring Approach would be a good idea, and, if the volume of traffic indicates, asking for advisories.

While you're still on the ramp at Newton:

You: Newton Traffic, Cherokee One Four Six One Tango at the terminal, taxiing to Runway One Three, Newton.

After engine runup:

You: Newton Traffic, Cherokee One Four Six One Tango taking One Three, southwest departure, Newton.

Off the ground and at 300 feet or so, you contact Fort Dodge Flight Service over the Newton VOR. In this instance, you transmit on 122.1 and receive on the VOR frequency of 112.5.

You: Fort Dodge Radio, Cherokee One Four Six One Tango listening Newton VOR.

FSS: *Cherokee One Four Six One Tango, Fort Dodge Radio, go ahead.*

You: Fort Dodge, Cherokee Six One Tango was off Newton at three five past the hour. Would you please open my flight plan to Kansas City Downtown?

FSS: *Cherokee Six One Tango, Roger. We show you off Newton at three five and will activate your flight plan to Kansas City. Des Moines altimeter three zero zero one.*

You: Roger. Thank you. Cherokee Six One Tango.

Although there is no ATA at Newton, one more local call is in order:

You: Newton Traffic, Cherokee Six One Tango now clear of the area to the southwest, Newton.

Monitoring Des Moines Approach as you pass through the ARSA's outer area, you call Minneapolis Center when clear of the area.

You: Minneapolis Center, Cherokee One Four Six One Tango.

Ctr: *Cherokee One Four Six One Tango, Minneapolis Center, go ahead.*

You: Center, Cherokee One Four Six One Tango is fifteen southwest of the Des Moines VOR at six thousand five hundred VFR to Kansas City Downtown via Victor One Three. Squawking one two zero zero. Request VFR advisories, if possible.

Ctr: *Cherokee Six One Tango, squawk zero five two three and ident.*

You: Cherokee Six One Tango squawking zero five two three.

Ctr: *Cherokee Six One Tango, radar contact. Report level at six thousand five hundred.*

Ctr: *Cherokee Six One Tango, Roger.*

There might or might not be further advisories from Center, depending on traffic. Whichever the case, you're soon over the Des Moines VOR and heading outbound on Victor 13.

After passing the Lamoni VOR, 53 miles out of Des Moines, you leave Minneapolis Center and enter the Kansas City Center area. As you cross the line:

Ctr: *Cherokee Six One Tango. Contact Kansas City Center now on one two seven point niner. Good day.*

You: Roger. One two seven point niner. Cherokee Six One Tango. Good day.

You: Kansas City Center, Cherokee One Four Six One Tango is with you at six thousand five hundred.

Ctr: *Cherokee Six One Tango, Roger. St. Joe altimeter two niner niner six.*

You: Roger, Cherokee Six One Tango.

About now is the time to tune the second radio to Kansas City on 112.6. With one VOR head tracking you outbound from Lamoni and the other inbound to Kansas City, you should be smack in the middle of V13.

When you're approximately due east of St. Joseph, Missouri, you spot some lightning not too far distant and just to the right of your course. This can be an omen of bad stuff, so you decide to check with Center. But wouldn't Center take it on itself to advise you of potential enroute weather problems? Not likely. In fact, it's unlikely. It's up to you to initiate the request for information:

You: Center, Cherokee Six One Tango. Request.

Ctr: *Cherokee Six One Tango, go ahead.*

You: Cherokee Six One Tango has lightning at one o'clock. Will my present course keep me clear of the storms?

Ctr: *Cherokee Six One Tango, affirmative. Scattered thunderstorms are moving northeast, but you should be past the area at your present ground speed.*

You: Roger. Thank you. Cherokee Six One Tango.

219

On the other hand, if a storm encounter seems likely, Center might offer this advice: "Cherokee Six One Tango, the storm activity is moving due east. Suggest right heading of two seven zero past St. Joe to Topeka and come in behind the weather." Keep in mind that such a suggestion is not a command. You're VFR, and have freedom as well as options.

Assuming the weather is not going to be a factor, Center will call you as you near the Kansas City TCA:

Ctr: *Cherokee Six One Tango, position one five miles north of the TCA. Contact Kansas City Approach on one one niner point zero.*

You: Roger, one one niner point zero. Thank you for your help. Cherokee Six One Tango.

Before contacting Approach, be sure you have monitored the Kansas City Downtown ATIS. Then, with the information clearly in mind, call Approach:

You: Kansas City Approach, Cherokee One Four Six One Tango is with you, level at six thousand five hundred with Foxtrot.

App: *Cherokee One Four Six One Tango, cleared into the TCA (Class B airspace), direct Kansas City VOR. Descend and maintain three thousand. Remain VFR at all times.*

You: Cherokee Six One Tango, Roger. Cleared into the TCA (Class B), direct Kansas City VOR, down to three thousand and remain VFR.

You: Approach, Cherokee Six One Tango level at three thousand.

App: *Roger, Six One Tango.*

Approach sees you nearing the VOR:

App: *Cherokee Six One Tango, descend to two thousand five hundred.*

You: Cherokee Six One Tango out of three for two thousand five hundred.

You: Approach, Cherokee Six One Tango level at two thousand five hundred.

App: *Roger, Six One Tango. Maintain heading and altitude.*

You: Will do. Six One Tango.

As you cross the VOR, only 9 miles from the airport:

App: *Cherokee Six One Tango, Downtown is niner miles at twelve o'clock. Advise when you have the airport in sight.*

You: Approach, Six One Tango has the airport.

App: *Roger, Six One Tango. Contact Downtown Tower on one three three point three.*

You: Will do. Cherokee Six One Tango.

You: Downtown Tower, Cherokee One Four Six One Tango is with you, level at two thousand five hundred.

Twr: *Cherokee One Four Six One Tango, continue straight in for Runway One Niner. Sequence later.*

You: Roger, straight in for One Niner, Cherokee Six One Tango.

Twr: *Cherokee Six One Tango, you'll be number two to land following a Citation on base.*

You: Roger. Cherokee Six One Tango has the Citation.

Twr: *Cherokee Six One Tango, Roger. Caution wake turbulence landing Citation. Cleared to land Runway One Niner.*

You: Cleared to land, Cherokee Six One Tango.

You watch the Citation touch down. To plan your landing because of the wake turbulence, you'd like a current wind reading.

You: Tower, Cherokee Six One Tango. Wind check.

Twr: *Cherokee Six One Tango, wind two one zero at one five.*

You: Cherokee Six One Tango.

Because you're coming in on Runway 19, these winds should blow the wake to the left of the runway, so you decide to land on the right, or upwind, side. You do so without difficulty and complete the rollout.

Twr: *Cherokee Six One Tango, contact Ground point niner when clear.*

You: Cherokee Six One Tango.

You: Downtown Ground, Cherokee One Four Six One Tango clear of One Niner. Taxi to the Flying Service.

GC: *Cherokee Six One Tango, taxi to the Flying Service.*

By radio when parked, or by phone, you close out the flight plan—and the trip concludes without incident.

CONCLUSION

The point of this cross-tower country was to illustrate the typical radio procedures when using Ground Control, Tower, Approach/Departure, Center, Flight Service, when operating within TCAs and ARSAs, and at unicom airports. The trip tried to encapsulate the more common phrases and phraseologies discussed in the various previous chapters.

Yes, there were several instances of what you might have considered needless repetition, but a cross-country involves repetition. Many of the same things are said to different agencies. Besides, repetition has a way of cementing habits in your mind and preventing errors and misunderstandings. Not every possible contact was included (as, for example, no call to Flight Watch), but many of the most common dialogues were re-created.

And, yes, some of the dialogues might seem a little stilted. As I stated early in the book, however, the examples are based on those illustrated in the *Airman's Information Manual (AIM)* and established as policy in the controller's *Air Traffic Control* manual, 7110.65G.

If you fly enough, you'll occasionally hear minor variations or local adaptations, particularly on the part of pilots. It's probably inevitable that a little slang or jargon, neither of which necessarily distorts the message, creeps in. Such liberties, however, don't conform to FAA procedures and are thus not acceptable. Consequently, I've tried to be as accurate and literal as possible in illustrating the approved communicating techniques—both in this theoretical cross-country and throughout the entire book.

In the process, I hope that the use of the radio and the services available to every pilot are just a bit clearer. If such is the case, perhaps some of the communicating concerns experienced by so many VFR pilots—both new and experienced—have been allayed at least a little.

15
A Final Word

Proper radio procedures are perhaps the most overlooked or underemphasized aspect in pilot training programs. Many budding pilots are never taught how to record and organize nav and com frequencies for easy reference and how to set up the frequencies in advance (assuming two navcoms are on board) to avoid excessive dialing at changeover points. They receive only the very basics of air-to-ground communications: what to say to whom and what to expect to hear in response. However, just as important as knowing what to say is how to say it.

Of course, there are exceptions to these criticisms. Some instructors like to teach communications and produce excellent pilots in the process. Their students sally forth into controlled areas with skill and confidence.

For those not privy to such training, it's a different story. Even many "experienced" pilots fear that they won't understand an instruction from a controller and that they will sound stupid over the air. The result is that those pilots avoid controlled airports and other FAA services available to them—services already paid for by their tax dollars. The fears are both natural and understandable for pilots who haven't been trained in radio techniques.

Everyone has qualms when they make those first tentative calls. Everyone has since screwed up a transmission at one time or another. So what? Mistakes in flying an airplane can be very fatal; mistakes over the air are usually no more than embarrassing—if that. If you learn what to say, how to say it, and aren't afraid to ask a controller to "say again" or "say more slowly" when you haven't understood, you'll find that your flights, local or otherwise, will be safer and more secure, and you'll have the confidence to venture into that tower-controlled airport that you perhaps have been avoiding.

If this book has clarified just a few of the areas in the radio communications process, it has met its objective. This project was undertaken because the need existed for a work of this nature. By so doing, I hope that any concerns have been put to rest and that your flying will be more enjoyable.

There are others in the air, however, who have no concerns and no doubts, but are languishing in unrecognized ignorance. They bust into TCAs and ARSAs unannounced; they monopolize the air with trivia; they hem, haw, mumble, and meander; their messages are unplanned, their transmissions disorganized. They take 4 minutes to communicate what the pro does in four seconds. Are they aware of their deficiencies? You know better. The air is for them, and let the rest take the hindmost . . . .. As is so often the case, those who need help the most are the last to recognize that need and ask for the help.

The sky is for all of us. For those who like to venture forth, do it with confidence and professionalism. Radio skills don't make better pilots at the controls, but they certainly add to competence and, quite logically, professionalism. Therein lies much of the joy of flying.

Abbreviations

The following is a glossary of various terms cited in this book. For a more complete glossary of pilot/controller radio communications terms, refer to the *Airman's Information Manual*, available by subscription from the U.S. Government Printing Office. Annual reprints are available from TAB Books, Blue Ridge Summit, PA 17294.

AAS—Airport Advisory Service
ADF—Automatic Direction Finder
A/FD—Airport/Facility Directory
AFSS—Automated Flight Service Station
agl—Above ground level
AIM—*Airman's Information Manual*
AIRMET—Airman's Meteorological Information
ALT—Transponder switch position to activate altitude-reporting equipment
AOPA—Aircraft Owners and Pilots Association
App—Approach Control
ARSA—Airport Radar Service Area
ARTCC—Air Route Traffic Control Center
ATA—Airport Traffic Area
ATC—Air Traffic Control
ATCRBS—Air Traffic Control Radar Beacon System
ATIS—Automatic Terminal Information Service
ATP—Air Transport Pilot
CAVU—Ceiling And Visibility Unlimited

CD—Clearance Delivery
Center—Air Route Traffic Control Center (*See* **ARTCC**)
CFI—Certified Flight Instructor
CFII—Certified Flight Instrument Instructor
com—Communications (Also, send-and-receive side of the radio, or navcom)
CT—Control Tower
CTAF—Common Traffic Advisory Frequency
CZ—Control Zone
Dep—Departure Control
DF—Direction-finding
DF fix—Direction-finding fix
DF steer—Direction-finding steer
DME—Distance measuring equipment
EFAS—Enroute Flight Advisory Service (Flight Watch)
ELAC—Enroute Low Altitude Chart
ELT—Emergency locator transmitter
ETA—Estimated time of arrival
ETD—Estimated time of departure
ETE—Estimated time enroute
FAA—Federal Aviation Administration
FARs—Federal Aviation Regulations
FBO—Fixed-base operator
FSS—Flight Service Station
GC—Ground Control
HIWAS—Hazardous Inflight Weather Advisory Service
ICAO—International Civil Aviation Organization
ID—Identification
IDENT—Transponder button that transmits assigned transponder code
"Ident"—Controller instruction to push IDENT button
IFR—Instrument Flight Rules
IMC—Instrument Meteorological Conditions
IPAI/DS—Identification-Position-Altitude-Intentions or Destination-Squawk
MEA—Minimum enroute altitude
MOA—Military Operations Area
MOCA—Minimum obstruction clearance altitude
MRA—Minimum reception area
msl—Mean sea level
MTR—Military training route
multicom—Nongovernment air-to-air radio communication facility
Nav—Navigation (Also the navigation side of the radio, or navcom)
Navcom—Radio with navigation and communication capabilities
NDB—Nondirectional beacon
NFCT—Nonfederal Control Tower

nm—Nautical mile

NOTAM—Notice to Airmen

NPRM—Notice of Proposed Rule-Making

PCA—Positive Control Area

PIREP—Pilot weather report

RAPCON—Military Radar Approach Control facility

RCO—Remote Communications Outlet

RF—Radio failure

SAR—Search and rescue

SBY—Standby—a transponder switch position

SIGMET—Significant meteorological information

sm—Statute mile

Squawk—Controller instruction to enter a specific number code in the transponder

Stage II—Radar service that provides advisories and sequencing to VFR aircraft

Stage III—Radar service that provides separation and sequencing to VFR aircraft

SVFR—Special Visual Flight Rules or Operations

TAC—Terminal Area Chart

TCA—Terminal Control Area

Tracab—Terminal radar control in the tower cab

Tracon—Terminal radar control

TRSA—Terminal Radar Service Area

TWEB—Transcribed Weather Broadcast

UHF—Ultra high frequency

unicom—Nongovernment air-to-ground radio communication facility

UTC—Coordinated Universal Time (Greenwich, Mean Time)

VFR—Visual Flight Rules

VHF—Very high frequency

VOR—Very high frequency omnidirectional range station (provides course guidance information)

VORTAC—A VOR station combined with the military TACAN distance information (provides course guidance plus nautical mile distance to the VORTAC station)

Wilco—Will comply

XFSS—Auxiliary Flight Service Station

Zulu—UTC time frequently cited as ". . . Zulu time"

Index

Other Bestsellers of Related Interest

BUSH FLYING—Steven Levi and Jim O'Meara

Survive in the air and on the ground—in the most treacherous conditions—with this practical guide. Focusing on the basics of flying over sparsely populated regions, it emphasizes the unique skills needed by bush pilots when flying in the mountains and cold weather. It explores every facet of this unpredictable brand of flight, and by doing so helps you avoid costly in-flight mistakes. 168 pages, 112 illustrations. Book No. 3462, $16.95 paperback only

UNDERSTANDING AERONAUTICAL CHARTS
—Terry T. Lankford

Filled with practical applications for beginning and veteran pilots, this book will show you how to plan your flights quickly, easily, and accurately. It covers all the charts you'll need for flight planning, including those for VFR, IFR, SID, STAR, loran, and helicopter flights. As you examine the criteria, purpose, and limitations of each chart, you'll learn the author's proven system for interpreting and using charts. 320 pages, 183 illustrations. Book No. 3844, $17.95 paperback, $26.95 hardcover

ABCs OF SAFE FLYING—3rd Edition
—David Frazier

This book gives you a wealth of flight safety information in a fun to read, easily digestible format. The author's anecdotal episodes as well as NTSB accident reports lend both humor and sobering reality to the text. Detailed photographs, maps, and illustrations ensure that you'll understand key concepts and techniques. If you want to make sure you have the right skills each time you fly, this book is your one-stop source. 192 pages, illustrations. Book No. 3757, $14.95 paperback, $22.95 hardcover

THE ART OF INSTRUMENT FLYING
—2nd Edition—J. R. Williams

". . . as complete and up-to-date as an instrument book can be." —*Aero* magazine

Williams has updated his comprehensive guide to include all elements of IFR flight—flight director, Loran-C, and Omega navigational systems. And, enroute, area, TCA, and SID/STAR reference charts reflect current designations. The first edition won the 1989 Best Technical Book award of the Western Region of the Aviation/Space Writers Association. 352 pages, 113 illustrations. Book No. 3654, $19.95 paperback, $31.95 hardcover

AVOIDING COMMON PILOT ERRORS:
An Air Traffic Controller's View—John Stewart

This essential reference—written from the controller's perspective—interprets the mistakes pilots often make when operating in controlled airspace. It cites situations frequently encountered by controllers that show how improper training, lack of preflight preparation, poor communication skills, and confusing regulations can lead to pilot mistakes. 240 pages, 32 illustrations. Book No. 2434, $16.95 paperback only

BE A BETTER PILOT:
Making the Right Decisions—Paul A. Craig

Why do good pilots sometimes make bad decisions? This book takes an in-depth look at the ways pilots make important preflight and in-flight decisions. And, it dispels the myths surrounding the pilot personality and provides straightforward solutions to poor decision-making and determines traits that pilots appear to share—traits that affect the way they approach situations. 240 pages, 76 illustrations. Book No. 3675, $15.95 paperback, $24.95 hardcover

CROSS-COUNTRY FLYING—3rd Edition
—Paul Garrison, Norval Kennedy,
and R. Randall Padfield

Establish and maintain sound flying habits with this classic cockpit reference. It includes revised information on Mode-C requirements, direct user access terminal usage (DUAT), LORAN-C navigation, hand-held transceivers, affordable moving maps, and over-water flying techniques. Plus, you'll find expanded coverage of survival equipment, TCAs, fuel management and conservation, mountain flying techniques, and off-airport landings. 328 pages, 148 illustrations. Book No. 3640, $19.95 paperback only

FLYING VFR IN MARGINAL WEATHER
—3rd Edition—Paul Garrison, Norval Kennedy,
and Daryl E. Murphy

Here's an invaluable guide for every VFR pilot who must make cross-country flying decisions in the face of uncertain weather. You'll find specific information on determining what marginal weather is—ceiling, visibility, wind, turbulence, precipitation, and temperature. You'll also learn how to deal with situations ranging from flying VFR "on top" or between cloud layers to coping with thunderstorms, wind shear, and dust devils, to using an interstate highway to find your way in featureless terrain. 224 pages, 87 illustrations. Book No. 3699, $16.95 paperback, $26.95 hardcover

GENERAL AVIATION LAW—Jerry A. Eichenberger

Although the regulatory burden that is part of flying sometimes seems overwhelming, it need not take the pleasure out of your flight time. Eichenberger provides an up-to-date survey of many aviation regulations, and gives you a solid understanding of FAA procedures and functions, airman ratings and maintenance certificates, the implications of aircraft ownership, and more. This book allows you to recognize legal problems before they result in FAA investigations and the potentially serious consequences. 240 pages. Book No. 3431, $16.95 paperback, $25.95 hardcover

GOOD TAKEOFFS AND GOOD LANDINGS
—2nd Edition—Joe Christy,
revised and updated by Ken George

This second edition is a complete guide to safe, precise takeoffs and landings. This updated edition includes new material on: obstructions to visibility, wind shear avoidance, unlighted night landings, and density altitude. You'll also find a recap of recent takeoff and landing mishaps and how to avoid them, expanded coverage of FARs, and information on the new recreational license. Special emphasis is placed on precision, and on safe practices that should become important habits. 208 pages, 76 illustrations. Book No. 3611, $15.95 paperback only

LEARNING TO FLY HELICOPTERS
—R. Randall Padfield

If you've always pictured yourself flying helicopters—whether you're a student just learning to fly or a fixed-wing pilot allured by the uniqueness of these machines—here's the guide you need. It's a conversational, often entertaining look at the principles and practices behind helicopter operation. It instructs beginners on everything from basic takeoffs and landings to advanced maneuvers in remote areas and high altitudes, standard emergency procedures, and the effect of human factors on flight safety. 368 pages, 210 illustrations. Book No. 3763, $19.95 paperback, $29.95 hardcover

THE PILOT'S AIR TRAFFIC CONTROL HANDBOOK—Paul E. Illman

This in-depth look at airspace system operations—told from the VFR pilot's point of view—focuses on the human elements of pilot and controller. An understanding of what one expects of the other, and the responsibilities that belong to each are examined—answering important questions often asked by VFR pilots. Direct quotes and suggestions from ATC personnel are highlighted. 240 pages, 88 illustrations. Book No. 2435, $16.95 paperback only

**THE PILOT'S GUIDE TO WEATHER REPORTS,
FORECASTS & FLIGHT PLANNING**
—Terry T. Lankford

Don't get caught in weather you're not prepared to handle. Learn how to use today's weather information services with this comprehensive guide. It shows you how to access weather services efficiently, translate briefings correctly, and apply reports and forecasts to specific preflight and in-flight situations to expand your margin of safety. 397 pages, 123 illustrations. Book No. 3582, $19.95 paperback, $29.95 hardcover